International Union of Geological Sciences
Commission on Tectonics

C. W. Passschier J. S. Myers A. Kröner

Field Geology of High-Grade Gneiss Terrains

With 101 Figures

Springer-Verlag
Berlin Heidelberg New York
London Paris Tokyo
Hong Kong Barcelona

Dr. Cees W. Passchier
Institute of Earth Sciences
Budapestlaan 4
NL-3508 TA Utrecht

Dr. John Stuart Myers
Geological Survey of Western Australia
1000 Plain Street
Perth, Western Australia 6004
Australia

Professor Dr. Alfred Kröner
Institute of Geosciences
University of Mainz
Saarstr. 21
D-6500 Mainz

Cover Illustration: Ancient sculpture of the Buddha entering Nirwana; in high-grade gneiss with tightly folded layering, folded neosome veins, and late pegmatite veins cutting the folds. Polonnaruwa, Sri Lanka

ISBN-13: 978-3-540-53053-4   e-ISBN-13: 978-3-642-76013-6

DOI: 10.1007/978-3-642-76013-6

© Springer-Verlag Berlin Heidelberg 1990

Softcover reprint of the hardcover 1st edition 1990

2132/3145-543210 — Printed on acid-free paper

Preface

Although there are numerous publications on the geology of high-grade gneiss terrains, few descriptions exist of how to map and carry out structural analysis in these terrains. Textbooks on structural geology concentrate on techniques applicable to low-grade terrains. Geologists who have no experience of mapping high-grade gneisses are often at a loss as to how to apply techniques to high-grade rocks that were developed for low to medium grade metamorphic terrains.

Any study of deep crustal processes and their development through time should begin with examination of the primary data source - outcrops of high-grade metamorphic terrains. We feel that the urge to apply advanced techniques of fabric analysis, petrology, geochemistry, isotope geochemistry and age determination to these rocks often results in brief sampling trips in which there is little, if any analysis of the structural and metamorphic history revealed by outcrop patterns. Many studies of the metamorphic petrology and geochemistry of high-grade gneiss terrains make ineffective use of available field data, often because the authors are unaware of structural complexities and of the ways to recognise and use them. This is unfortunate, because much data can be collected in the field at minimal cost that cannot easily, if at all, be obtained from material in the laboratory. The primary igneous or sedimentary nature of a rock, the relative age of intrusive veins, and the sequence of deformation that they underwent, can usually best be determined by straightforward observation in the field. Failure to collect such data will greatly diminish the value of geochronological and geochemical results.

In this manual we outline some of the more commonly seen structural features of high-grade gneiss terrains, and suggest how they can be analysed and interpreted, and which sampling techniques and isotopic age determinations can give the best results. We do not intend to give a *foolproof* field-guide to high-grade gneisses, but rather to provide thought-provoking guidelines for solving structural problems. We assume that the reader is familiar with the common terminology of structural geology as used in textbooks such as Hobbs et al. (1976), Ramsay and Huber (1983, 1987) and Suppe (1985). Some of the less well known terminology is explained in footnotes.

This manual results from an IUGS-AGID workshop on the "evolution of high-grade gneiss terrains", held at Kandy, Sri Lanka in August 1987. We are grateful to the Commission on Tectonics (IUGS) for its encouragement to publish this work. We wish to thank G. Clarke, P. G. Cooray, P. Dirks, S. Harley,

VI

L. Kriegsman, C. Mayer, M. Raith, W. Schreyer, T. Senior, C. Wilson and C. Simpson for their contribution to this manual and P. Jaeckel for assistance in preparing the text for printing. A. K. acknowledges financial support from the Deutsche Forschungsgemeinschaft for fieldwork in Sri Lanka. C.W.P. acknowledges financial support by the Schürmann Fund. J.S.M. thanks the directors of the Geological Surveys of Greenland and Western Australia for experience gained whilst working for these Surveys, and publishes with their permission.

Utrecht - Perth - Mainz - June 1990

Cees Passchier
John Myers
Alfred Kröner

Contents

VIII

1 Introduction

High-grade[1] gneisses make up a significant part of the continental crust exposed in orogenic belts and in Precambrian cratons. A lot of research has been focussed on high-grade gneiss terrains because they contain a large proportion of the Earth's mineral wealth. Some of these terrains are considered to represent exhumed segments of the lower continental crust and, as such, give valuable insight into processes of crustal genesis.

If we examine any high-grade gneiss terrain in detail, we will usually see a great variety of rock types in layers and lenses with complex geometry and interrelationships. Gneisses with various compositions and internal structures usually occur together. They are generally quartzo-feldspathic rocks with various amounts of biotite, hornblende, diopside, hypersthene, garnet, Al-silicates, cordierite, zircon and opaque minerals. Minor components of gneiss terrains include quartzites, pelites, calc-silicate rocks or marbles, metabasic and ultra-mafic rocks. The internal structure and interrelationship of rock types in high-grade gneiss terrains are usually complex, because of a long and eventful history. Some of these terrains may have originated as sedimentary rocks that spent a considerable time at deep crustal levels before being uplifted and eroded. Intrusive rocks may have been repeatedly emplaced during several episodes, and the original metasedimentary rocks may consequently form only a minor fraction of the total rock volume (Fig. 1.1). In addition, several phases of ductile and brittle deformation may have complicated the structure of the rocks.

The complexity of the structure seen in a gneiss terrain is, however, not simply an additional effect of all the phases of metamorphism, intrusion and deformation. A gneiss may not necessarily reveal all the aspects of its history clearly, because younger deformation processes tend to flatten older structures, thereby obscuring or even erasing them. Recrystallisation and partial melting are other mechanisms that blur or obliterate delicate older structures and metamorphic assemblages. The development of the internal structure of a gneiss terrain is a continuous competition between factors which tend to destroy evidence of older events, and compound visible geological history (Fig. 1.1). Failure to recognise and correctly assess the nature and complexity of a gneiss terrain in the

---

[1] High-grade metamorphic conditions coincide with those of the uppermost amphibolite and granulite facies. The word 'granulite' is commonly used to indicate a high-grade metamorphic rock, e.g. a felsic or mafic granulite.

field could lead to a completely wrong interpretation of the geological history, and to analytical work that would be of little value.

This manual presents a broad outline of the most practical methods to attack geological problems in high-grade gneiss terrains. We restrict ourselves to the field aspects of such studies - to the actual collection of field data and the choice

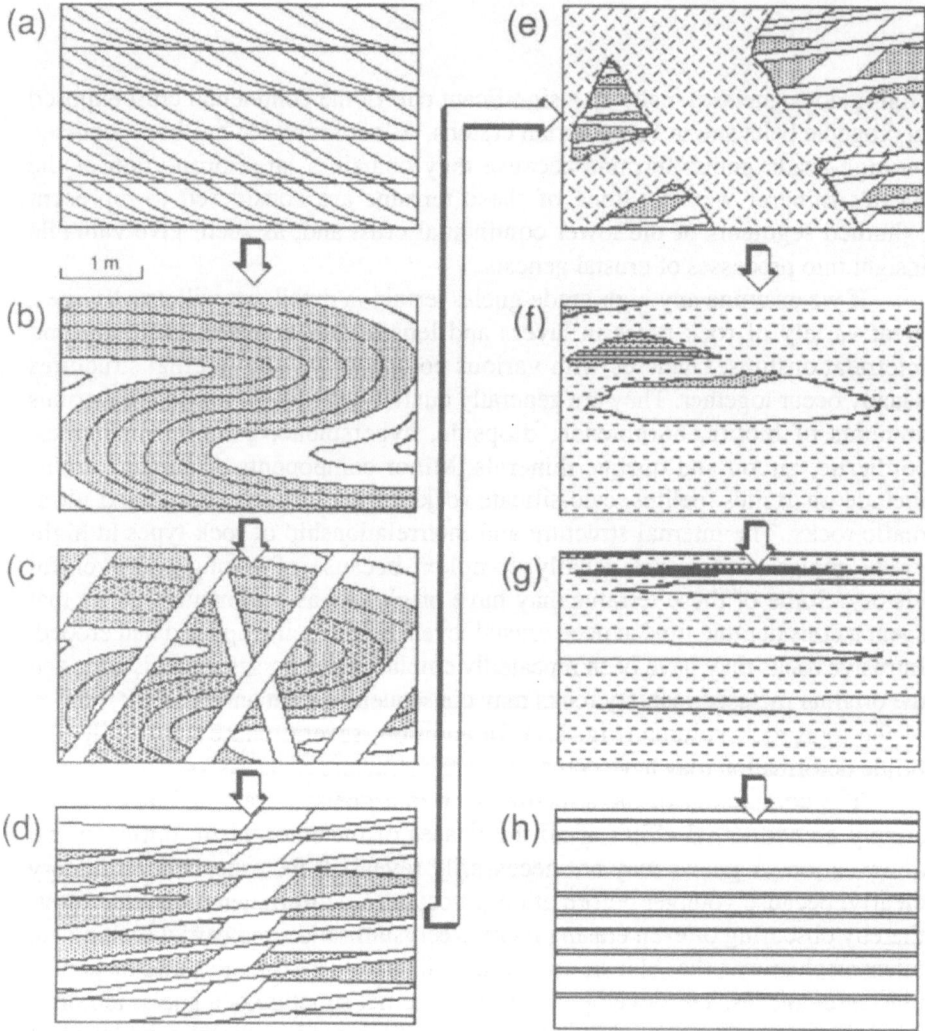

Fig. 1.1. A homogeneous, apparently undeformed layered gneiss (h) may have been derived by the sequence of events a-h: starting as a sediment (a), followed by phases of deformation (a-b, c-d, e-h) and intrusion (b-c, d-e). The intensity of deformation and the amount of igneous material involved is much greater than would be expected from the final result. The original volume of sedimentary rocks (in grey) forms only a small percentage of the final gneiss.

of sites for sampling. One of the principal aims is to help workers in gneiss terrains to recognise the effects of repeated complex deformation in the field. The manual is subdivided as follows:

Chapter 2 explains some of the methods of mapping in gneiss terrains; it gives examples of troubleshooting and describes how to work out a structural and metamorphic sequence; it also deals with sampling techniques.

Chapter 3 outlines current understanding of the development of the most common structures and fabric elements of gneiss terrains. It provides a necessary background for the correct interpretation of structures in the field.

Chapter 4 deals with the interpretation of structures in outcrop, and explains a selection of commonly encountered problems.

Chapter 5 gives a short outline of metamorphic mineral assemblages in high-grade gneiss terrains, as far as these can be recognised in the field, and deals with aspects of the metamorphic evolution of such terrains.

Chapter 6 treats some of the most useful aspects of geochemistry, isotope geochemistry and geochronology applicable to high-grade gneiss terrains.

Chapter 7 outlines some recent ideas on the genesis of high-grade gneiss terrains.

Chapter 8 is a problem section with exercises.

2 Mapping in Gneiss Terrains

2.1 Introduction

Whatever the purpose of working in gneiss terrains, a prerequisite for success is a thorough understanding of the geometry and relative age relations of rock units and the sequence of deformation and metamorphism. Because of the three-dimensional complexity of deformation and intrusion relations in gneiss terrains, it is in most cases impossible to understand relations by visiting a single outcrop, or even by means of a small number of transects through the area under consideration. If no detailed geological maps are present, or if the subject of study is insufficiently dealt with on available maps, it will be necessary to map an area in some detail.

There seems to be a growing tendency in geology to regard mapping as something of a former age which has no place in the modern world of micro-probes, mass spectrometers and mainframe computers. We believe, however, that 'mapping' is an integral and fundamental part of any research project involving sampling of 'real rocks'. With 'mapping' we do not exclusively mean 'to produce a map'; we mean the whole spectrum of field activities involving outcrop analysis and assessment of metamorphic conditions; establishment of a regional scheme of events and any local deviations or regional gradients; and establishment of the regional 3D geometric distribution of rock units, major structures and certain minerals. These are best displayed in profiles, block diagrams, tables and, of course, on maps. These activities provide a solid basis for laboratory work on collected material and help to extract the largest possible amount of data from the study area, avoiding errors resulting from poor or insufficient observations.

2.2 General Problems

Training exercises for mapping are usually carried out in non- or low-grade metamorphic terrains with clearly recognisable stratigraphy and do not provide sufficient experience for the problems encountered in mapping high-grade gneiss terrains. Well known problems are:

(1) Strong deformation and new growth of minerals may blur or destroy the sedimentary or igneous fabric[1] of rocks, and obscure the initial stratigraphic sequence. Repetition or omission of parts of an original stratigraphy by isoclinal folds, shear zones or thrusts may be very difficult to detect.

(2) Classical overprinting criteria such as crenulated foliations and small-scale refolded folds are less commonly seen in high-grade gneiss terrains than in medium- to low-grade areas. In gneisses, overprinting criteria involving minor intrusions, partial melting relations and shear zones must also be used.

(3) Most fold structures are not cylindrical in gneiss terrains; complex three-dimensional interference patterns of folds and shear zones are widely developed.

(4) Absolute dating of high-grade rocks cannot be undertaken in the field; it is virtually restricted to radiometric dating of intrusive or volcanic rock units. Fossils are generally distorted and difficult to classify, absent, or are destroyed by deformation and metamorphic growth of minerals.

(5) Measurement of the orientation of small-scale fabric elements is difficult in many gneiss terrains since outcrop surfaces are often smooth.

2.3 Working Method

The working method in high-grade gneiss terrains depends on the actual purpose of the fieldwork; if the regional tectonic evolution of a terrain is the subject of study, extensive mapping may be required, involving all aspects treated in this manual; if the subject is more restricted, such as the development of a specific fabric element in pegmatites or the age of a specific intrusion, small-scale mapping of a selected area or just a quarry face may be sufficient. In most cases, several outcrops will have to be visited; each outcrop must be analysed and the data be used to determine a local scheme of events; the map is a tool to order part of the data and interpret the origin of certain features.

2.4 Outcrop Analysis

Of all the features visible in outcrop, one must concentrate on the main fabric elements; their orientation, symmetry and relative age. For each major outcrop, a sequence of deformation and intrusion events must be determined and, if possible, stable mineral assemblages (Section 5.1) for each event be established. Sketches (Fig. 2.1) must be made of features in outcrop, as they define the

---

[1] We use the term 'fabric' as in Hobbs et al. (1976), i.e. 'the complete spatial and geo-metric configuration of all those components that make up the rock. It covers terms such as texture, structure and preferred orientation and so is an all-encompassing term that describes the shapes and characters of individual parts of a rock mass and the way in which these parts are distributed and oriented in space'.

6

geometries of structures easier and more concisely than words or numbers. Such sketches should be oriented and marked with a scale, and the orientation of the outcrop surface must be noted (Fig. 2.1).

The relative age of features seen in outcrop is often difficult to establish. If this is the case, one must record the various possible interpretations and search for similar outcrops in which the situation may be more obvious. If strong deformation is the problem, a visit to an adjacent less deformed area, or even a small, little-deformed lens may solve the problem. Many gneiss terrains are well exposed; in such cases, one must first identify and study the least deformed areas where any surviving primary features and the oldest overprinting and intersection relations will be best preserved. Further details on the interpretation of features in outcrop can be found in Chapters 4 and 5.

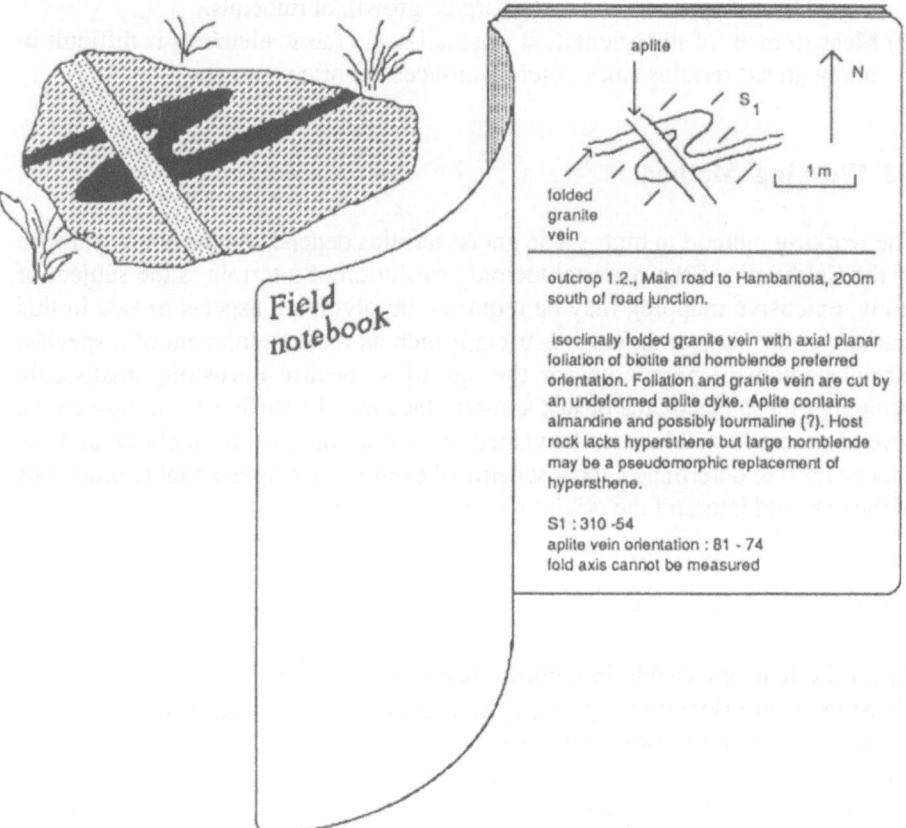

aplite

S₁

N

folded
granite
vein

1 m

outcrop 1.2., Main road to Hambantota, 200m south of road junction.

isoclinally folded granite vein with axial planar foliation of biotite and hornblende preferred orientation. Foliation and granite vein are cut by an undeformed aplite dyke. Aplite contains almandine and possibly tourmaline (?). Host rock lacks hypersthene but large hornblende may be a pseudomorphic replacement of hypersthene.

S1 : 310 -54
aplite vein orientation : 81 - 74
fold axis cannot be measured

Field
notebook

Fig. 2.1. Fabric elements visible in outcrop must be recorded by short descriptions and by sketches.

2.5 What to Map

With the help of this manual it should be possible to reconstruct a sequence of events from an outcrop. In most cases, however, the significance of individual events and the regional sequence can only be established by mapping a larger area.

Gneiss terrains are potentially some of the most complicated areas to map. They generally lack distinctive stratigraphy or marker horizons. The nature and intensity of tectonic and metamorphic modifications of primary features may be very heterogeneous. Rock units may occur as thin, discontinuous lenses which are too small and too numerous to plot on a conventional lithologic map. In addition the rocks may be cut by intrusions or shear zones, and folded into geo-metrically complex structures. There may be rapid transitions between little deformed and intensely deformed rocks, and between high-grade and medium-grade metamorphic rocks derived from the former by retrogression.

One of the first questions raised by anyone working in high-grade gneiss terrains for the first time is what should I map?' Some of the most obvious features to map may be the outline of large intrusions and the traces of major shear zones, but these generally reflect the geometry of relatively late events. To understand the earlier geological history one needs to look more closely at the superficially monotonous gneiss itself.

It is important to realise that, in most gneiss terrains, primary stratigraphy is strongly deformed and converted into a tectonic layering, and that many gneiss terrains are dominated by deformed igneous rocks of different ages. Therefore, conventional principles of stratigraphy cannot be applied in most gneiss terrains. In many cases, mappable horizons may be trains of distinct kinds of inclusions or veins that can be followed over great distances. Such features form useful tectono-stratigraphic marker horizons (Myers, 1971; 1981) by which large-scale structures can be recognised. For instance, trains of amphibolite fragments which occur as relicts in intrusive granitoid gneiss have been exten-sively mapped and described from the Archaean gneiss complex of southwest Greenland (Fig. 2.2; Bridgwater et al., 1976). The mapping of trains of inclu-sions or other marker horizons will indicate whether the main gneissose layering is parallel to primary stratigraphy or is a tectonic fabric, axial planar to major folds.

Pegmatite veins are an abundant feature of many gneiss terrains and provide small-scale mappable features. Different kinds and generations of pegmatite veins must be distinguished and their orientations and tectonic structure recorded. It is best to record all these minor details directly on the face of the map or air photograph. If, however, the scale of mapping is too large for large numbers of pegmatite veins to be recorded from a single outcrop, then the information must be set out in a sketch map of the outcrop in a notebook. At a later stage, overlay maps must be produced showing the distribution, orientation and structure of different generations of pegmatite veins, in the same way as maps can be produced showing structures of different tectonic episodes.

8

2.6 Types of Maps

Basic mapping techniques such as the use and interpretation of maps and air photographs, measurement of the strike and dip of planar structures, and direction and plunge of lineations are common to all kinds of geological terrains and so are not described here. General introductions to geological mapping can be found in textbooks such as Hobbs et al. (1976); Barnes (1981) and Fry (1984).

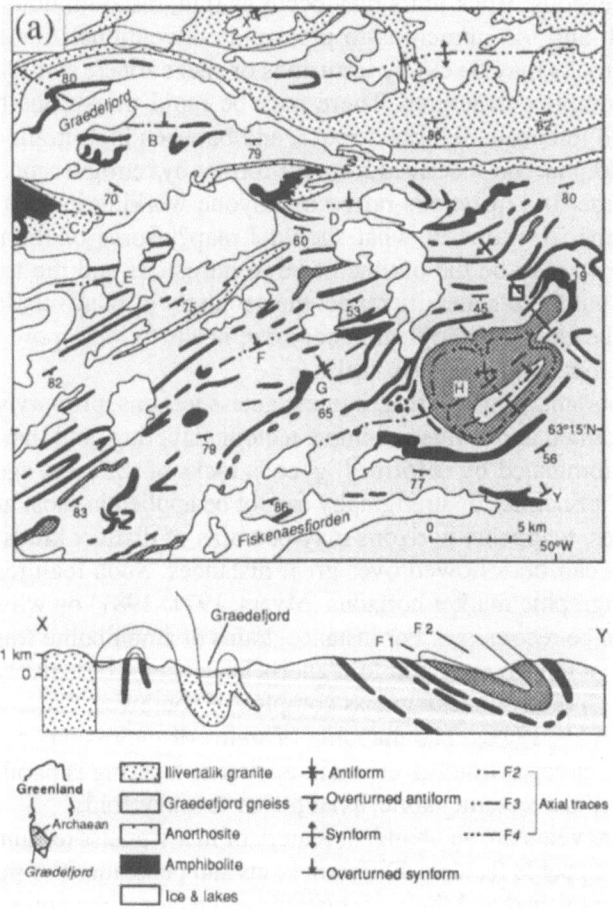

Fig. 2.2a. Conventional geological map and profile of the Grædefjord area in the Archaean gneiss complex of SW Greenland. Square indicates location of Fig 2.3. After Myers (1978).

Data on rock type, orientation of structural elements and metamorphic conditions (including the outcrop number) can be plotted on a conventional geological map. Fig.2.2a shows such a map and accompanying profile for an area in SW-Greenland (Myers, 1978). Many other kinds of maps can be constructed for specific purposes, e.g. isograd maps, maps showing the distribution of boudins of a particular lithology in a gneiss terrain, and finite strain maps. Figure 2.2b

9

shows a finite strain map for the area in Fig. 2.2a, based on the change in shape of amphibolite fragments in the gneiss. This map serves to show the geometry and intensity of large-scale deformation in the area. Because complex overprinting relations are best preserved in little deformed areas, the map could be used to select such domains for detailed study.

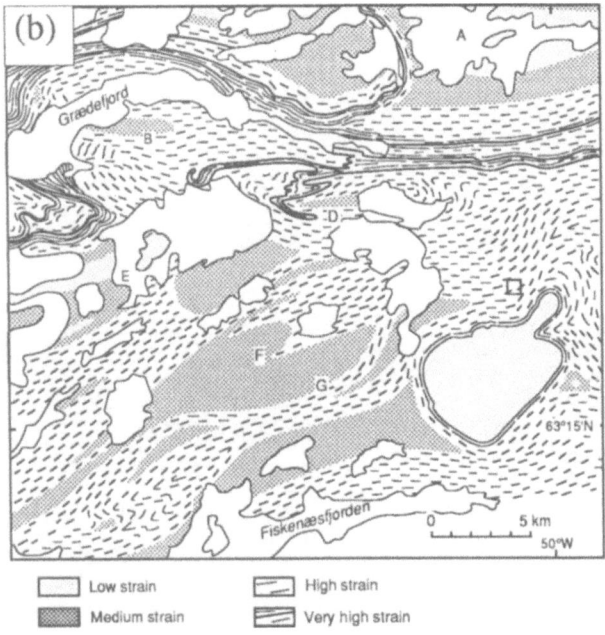

Fig. 2.2b. Finite strain map of the Grædefjord area in the Archaean gneiss complex of SW Greenland. Square indicates location of Fig 2.3.

In many high-grade gneiss terrains there are few rock units which can be followed over a large distance. It can be difficult to interpret the results of mapping in such terrains. Figure 2.3a shows a small-scale map of an area of homogeneous gneiss such as may be found in the area of Fig. 2.2. Such a map can provide sufficient data to unravel the local sequence of events and geometric complexity, if used in combination with sketches in a notebook. However, in practice it is difficult to manage the available data. A solution to this problem, often·applied in lithologically monotonous gneiss terrains, is the construction of *form-surface maps* (Fig. 2.3b; Hobbs et al., 1976). Form-surface maps show the trace of certain foliations, outlining major structures which would otherwise be obscured in the mass of dip and strike symbols. Outcrop boundaries can be drawn in order to distinguish between observations and interpretation. Important features from the outcrop are marked directly on the map, such as minor asymmetric folds (projected to one standard plane and with the correct vergence[1]), intersecting foliations, the trend of a certain type of intrusive vein, or trains of boudins of a particular composition. Form-surface maps can be of considerable value in unravelling the structural complexity of a gneiss terrain, since most of the available information can be brought together on a single map in a condensed and easily readable manner.

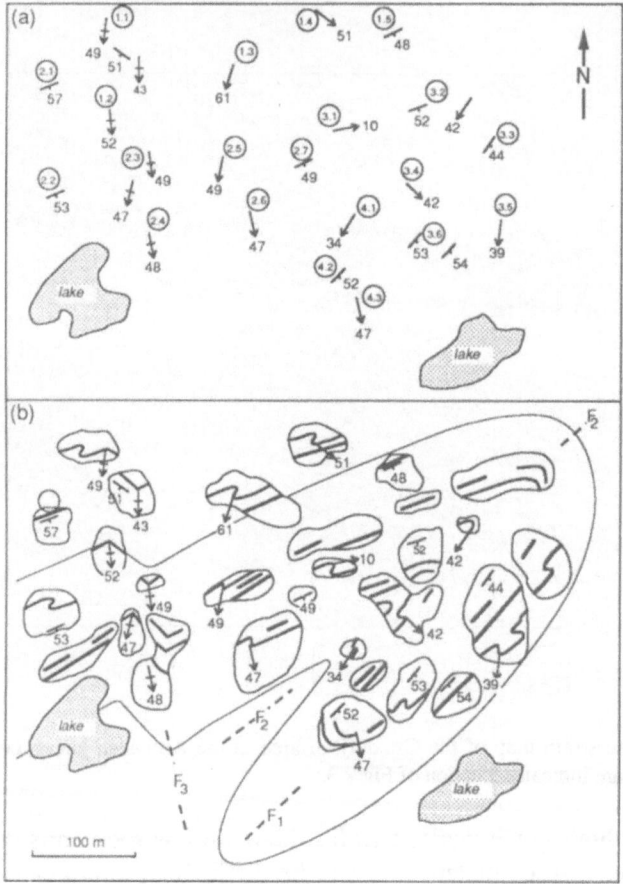

Fig. 2.3. (a) Imaginary conventional geological map representing a domain of homogeneous gneiss in the Grædefjord area, Greenland. Outcrop numbers are shown; (b) the same area shown as a form-surface map with outlines of major outcrops. In the outcrops, the trace of the main foliation is shown as a solid black line and the vergence of parasitic folds is indicated. Thin lines outside the outcrops represent the trace of the main foliation in the area, as interpreted from the data in the outcrops. The vergence of parasitic folds indicates that three phases of deformation are represented on the map by macroscopic folds. This overall structure would be difficult to deduce from (a), even if the same data were available in notebook sketches. Outcrop numbers are omitted from map (b) for clarity.

---

[1] Vergence as used here describes the asymmetry of parasitic folds on the limbs of a larger fold, as seen on the map; parasitic folds can have sinistral or dextral vergence.

2.7 Scheme of Events

It is important to start building a regional scheme of events as soon as mapping commences, and to use and develop this scheme as a 'working model' as the mapping progresses. An established sequence of deformation events, intrusions, and metamorphic events must be checked at each outcrop, and gradually expanded and modified according to new observations. Figure 2.4 shows an imaginary sequence of notebook sketches based on a series of outcrops visited, and the updated scheme of events for each newly discovered relation. The structures and sequence of events are imaginary but based on the real sequence of events in the area of Fig. 2.2 (Myers, 1978). If relations are found that are in conflict with part of the scheme of events, a careful analysis must be made of their nature and of consequences for the rest of the scheme. They may reflect a local deviation from the regional pattern, or one of the basic assumptions or the interpretation of a fabric element may be wrong. Such inconsistencies usually mean that outcrops were not investigated in sufficient detail. The 'working model' can be used to build up a field scheme of events that is completed with the mapping. Mapping is finished when no new boundaries or structures can be traced and no further important additions can be made to the scheme of events when new outcrops are visited.

2.8 Sampling

Sampling for microstructural analysis and microprobe work should be undertaken during mapping, in order to obtain a collection which can be used to refine and check the final field scheme of events. Before a domain of rock is sampled, one should be able to answer the following questions: what is the probable source of the material? When was it emplaced in its regional setting? What is its relative age with respect to adjacent domains or veins? Is it deformed, and what is the relative age of the deformation with respect to the mineral assemblage in the rock? Can the deformation features be interpreted in terms of extensional or constrictional shear zone development, large-scale folding, or polyphase deformation? Without careful notes of all these relationships, samples lose much of their value. A careful analysis of a mineral assemblage in a sample can be worthless if the relative age during which an assemblage was stable is unknown. No reliable pressure-temperature paths can be determined in such a case, and nothing can be said of the development of the domain with time. If present, shear zones, boudin necks and fold closures must be sampled, as re-equilibration of mineral assemblages generally occurs in those locations; such samples can help to determine metamorphic conditions during specific deformation events (Chapter 5).

12

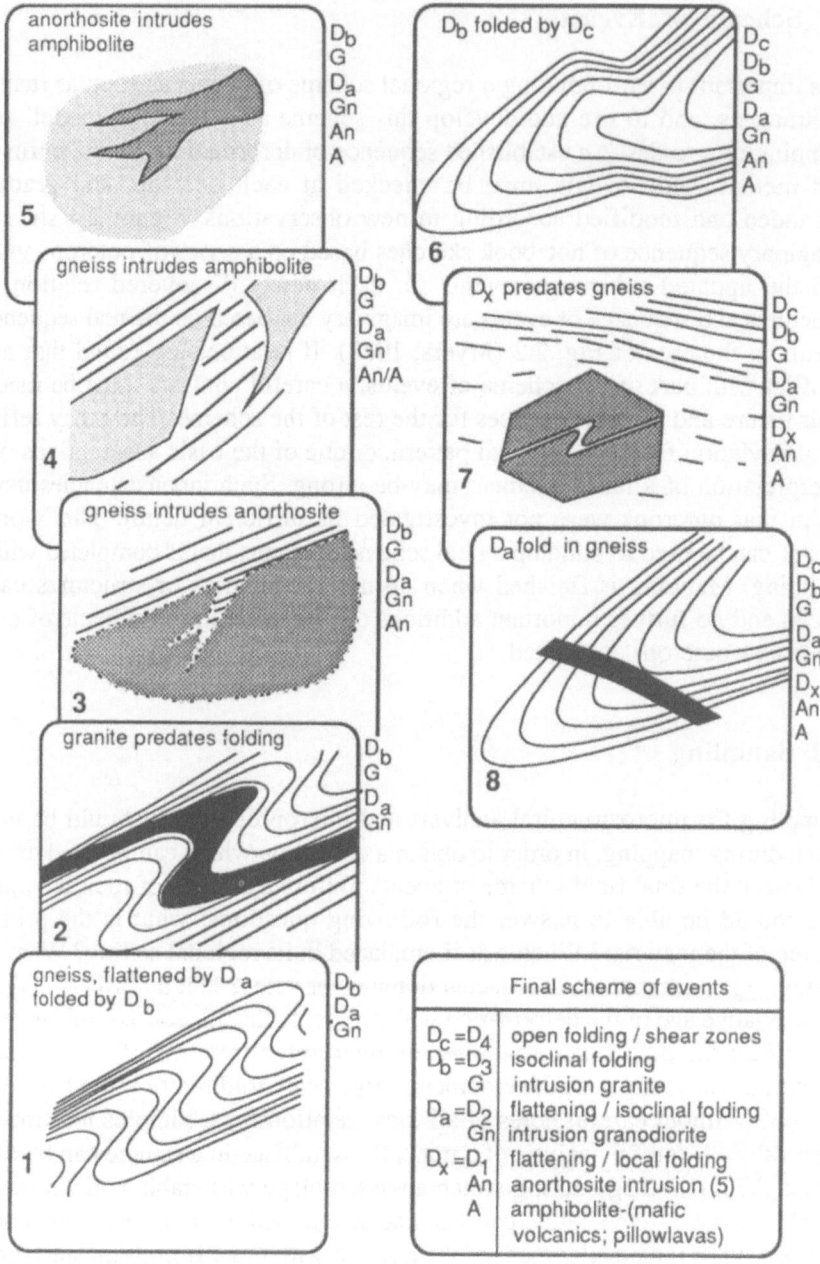

Fig. 2.4. Imaginary series of notebook pages representing a series of outcrops encountered in the order 1-8. The series illustrates the way in which a scheme of events is constructed and modified as more data becomes available. Phases of deformation can be abbreviated as Da, Db etc. until the scheme has been completed, after which the usual sequence of D1, D2 etc. can be adopted.

exact provenance of a sample must be described and sketched in sufficient detail to be able to place it in the scheme of events. Was pegmatite A or B sampled in an outcrop with cross-cutting veins? What is the relative age of the lineation in the sample? Questions like these should be easy to answer back home from the notebook.

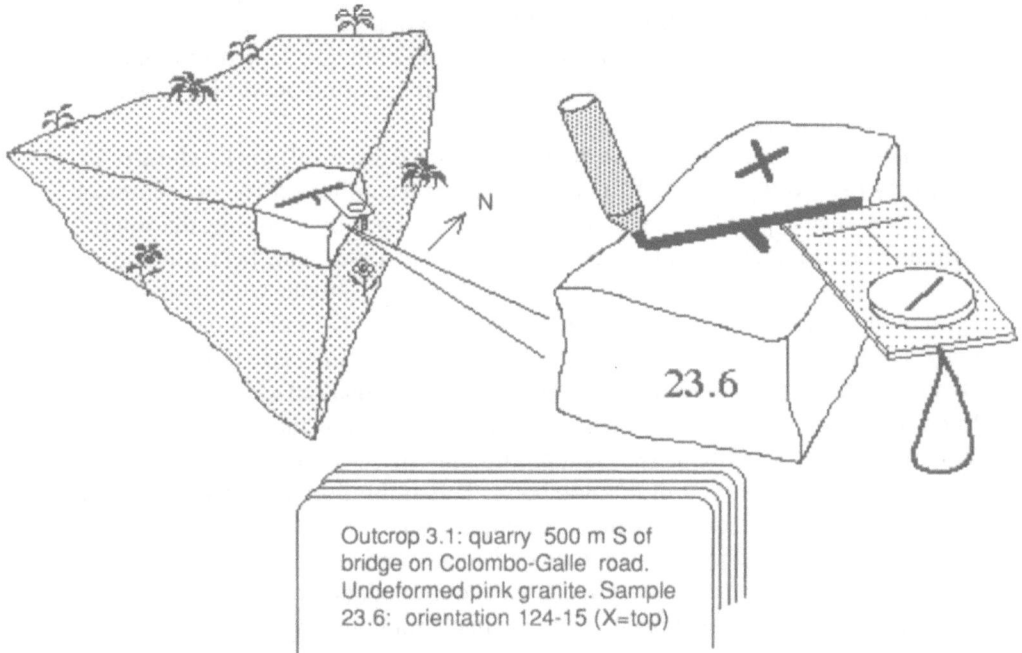

Outcrop 3.1: quarry 500 m S of bridge on Colombo-Galle road. Undeformed pink granite. Sample 23.6: orientation 124-15 (X=top)

Fig. 2.5. Illustration of how to orient a sample; after detaching the sample, it is fitted back into its original position, and a (relatively) planar surface is selected for measurement. The orientation of this surface is marked with a dip and strike symbol. An extra mark (e.g. X for the top) is needed to indicate whether the measured surface is the top or bottom of the sample. The sample is then numbered, and the orientation of the marked surface is entered into the notebook. In this case dip direction (124) and dip (15) have been measured. A sketch of the setting where the sample is taken may be useful if the local structure is complex.

Sampling for age determinations as well as isotopic and geochemical research must ideally be undertaken after the local scheme of events has been established, in order to avoid sampling 'the wrong rocks'. Geochemical work usually requires the collection of large samples (~1-20 kg) because many high-grade rocks are coarse-grained and exhibit compositional layering. Since the main aim of sampling is to obtain a specimen representative of the rock unit as a whole, sample size strongly depends on the grain size and the visible degree of mineralogical inhomogeneity. For example, it may be sufficient to collect a ~1 kg specimen from a fine-grained metabasalt or a mafic dyke for geochemical work, while a coarse grained charnockite, a migmatite or a coarsely layered gneiss require specimens weighting 20 kg or more.

If possible, sample only fresh, unweathered material (which can be a problem in some gneiss terrains). If P-T conditions must be determined, samples from metabasic rocks, calcsilicates and pelites are more useful than those from quartzo-feldspathic gneisses. In pelites and metabasic rocks, one must sample material with a relatively high proportion of Mg-Fe-Si minerals such as biotite, amphiboles, pyroxenes, cordierite and garnet. Reactions involving these minerals are relatively well known, and many geothermometers and geobarometers can be used on them (see Chapter 5). Where retrograde re-equilibration of minerals is suspected, it can be advantageous to sample both coarse and fine grained material from the same lithology. Re-equilibration has often proceeded further in fine grained than in coarse grained material.

All samples, not only those taken for microstructural work, **should be properly oriented in the field** (Fig. 2.5). Samples collected for geochemical work may reveal problems which can only be solved by studying the microstructure; unoriented samples are useless in that case. The best method of orienting samples is to mark a planar surface of the sample with a strike-dip symbol showing its orientation in outcrop (Fig. 2.5). On horizontal surfaces, mark North with an arrow. This still leaves two possible orientations of the sample, so the top or bottom of the sample has to be marked as well (Fig. 2.5). In complex outcrops, it is very useful to make a sketch of the overall structure showing where the sample was taken. Dip and strike values, and the sample number should be marked in the notebook. This minimises problems that may arise if lettering wears off the samples during transport.

When sampling a layer, shear zone or fault of particular interest, it is a good idea to sample the host rock as well; chemical differentiation can only be detected if sufficient samples of the host rock are available for comparison.

3 Fabric Development in Gneiss Terrains

3.1 Introduction

This chapter describes some aspects of the development of fabrics in gneiss terrains. It provides essential background information that should be read before attempting to map a gneiss terrain. Inevitably, the interpretations are 'state of the art' and not necessarily the absolute answer. Additional information can be found in the cited literature and various specialised journals such as the *Journal of Structural Geology*, the *Journal of Metamorphic Geology*, *Tectonophysics*, *Tectonics* and *Precambrian Research*.

3.2 The Geometry of Ductile Flow in Rocks

3.2.1 Coaxial or Non-Coaxial Flow

At medium to high metamorphic grade, rocks mainly deform in a ductile manner; they change shape without developing sharp, discrete macroscopic fractures. Even without the development of fractures, deformation can be of different types and can be homogeneous or heterogeneous on any scale (Fig. 3.1).

On a microscopic scale, progressive deformation is governed by complex grain-scale deformation mechanisms[1] (Poirier, 1985; Behrmann & Mainprice, 1987), but on the larger scale of an outcrop it can be described by continuum flow such as in a liquid (Fig. 3.1). At every instant during progressive deformation, the 'instantaneous' flow can be different from that at other instants. In this manual, 'flow' always means instantaneous flow. Flow is defined by the movement direction (at a particular instant) of all particles in the volume of material under consideration. Such an unpractical description can be simplified if we subdivide the volume of material into sufficiently small domains. We can then consider the flow in each domain as homogeneous, that is identical from place to place within that volume (even if this cannot usually be done in practice, it serves as a useful mental hypothesis). In such a small volume of material, flow at any instant can be described by the displacement vectors of a set of particles,

---

[1] These mechanisms include migration of lattice defects (dislocations, vacancies) in crystals, diffusion of material through the crystal lattice and along grain boundaries. We use the term 'crystalplastic deformation' for all these mechanisms.

16

or alternatively by the stretching rates[1] and angular velocities of imaginary lines of particles in the material ('material lines'; Fig. 3.2).

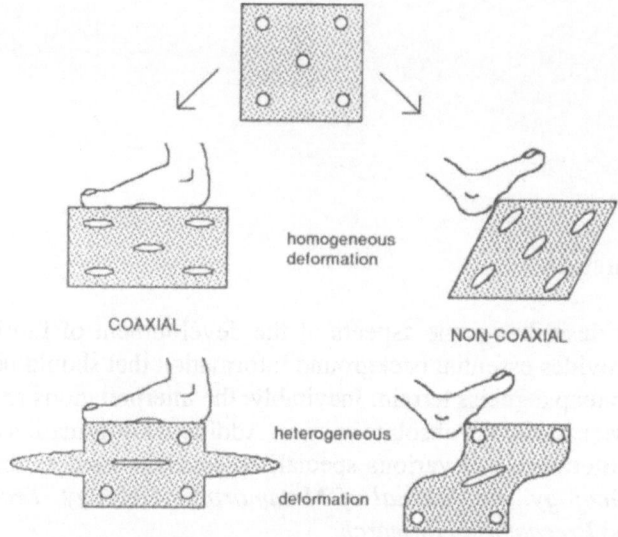

Fig. 3.1. How does a volume of rock deform in high-grade metamorphic conditions? Deformation can be homogeneous or inhomogeneous on the scale of observation, and may develop by coaxial or non-coaxial flow (see text).

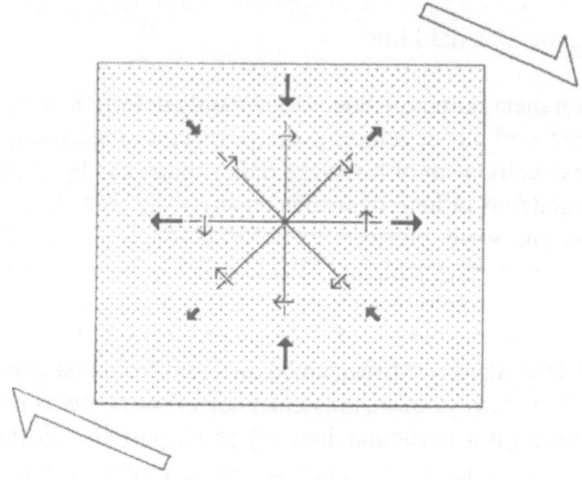

Fig. 3.2. Homogeneous progressive deformation or flow in a volume of rock can best be studied from the rotation and stretching behaviour of imaginary lines 'in' the material. Open arrows outside the volume represent external deformation which imposes a deformation rate on the volume of material considered; barbed and solid arrows in the volume indicate directions of angular velocity and stretching rate of material lines, respectively.

---

[1] Stretching rate is the rate of change in length of a line (new length divided by old length) per second.

Homogeneous flow is relatively simple and regular (Fig. 3.3, curves at right). Maximum and minimum stretching rates are realised along orthogonal spatial lines; these lines are known as the 'principal stretching axes' (PSA) of the flow (Passchier, 1986b; 1987). The summed angular velocity of two orthogonal lines in a two-dimensional flow is known as the flow vorticity (Means et al., 1980). The angular velocity of material lines which coincide with PSA can be used to classify flow types. We can distinguish two basic types, here for simplicity explained for a two-dimensional situation (Fig. 3.3; Lister & Williams, 1983):

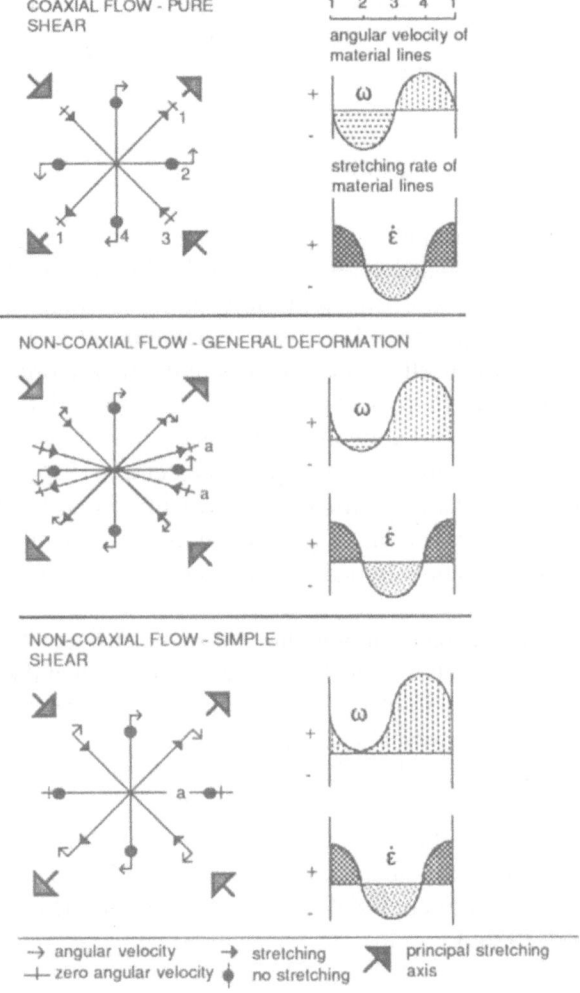

Fig. 3.3. This diagram illustrates flow types by the pattern of angular velocities (ω) and stretching rates (ε) of the material lines shown at left. These patterns are always regular (compare graphs at right), but the symmetry of the ω-graph defines different flow types, of which 'pure shear' and 'simple shear' are the end members. Important lines mentioned in the text are the axes of maximum and minimum stretching rate ('principal stretching axes' -PSA), and the lines of zero angular velocity (a).

(1) material lines that coincide with PSA of homogeneous flow do not rotate (coaxial or 'pure shear' flow). In this case the flow vorticity is zero. If this type of flow is active during a certain period (progressive deformation), the same material lines act as PSA all the time, while all other material lines rotate (coaxial progressive deformation).

(2) material lines that coincide with PSA of homogeneous flow do rotate (non-coaxial flow). In this case flow vorticity is a finite number. Different material lines act as PSA from time to time during 'non-coaxial' progressive deformation. One or two other lines, however, (a in Fig. 3.3) may remain stationary. 'Simple shear' is an endmember of the possible types of non-coaxial progressive deformation. In simple shear, there is only one line which does not rotate. For three-dimensional simple shear this is a surface known as the 'flow plane'.

3.2.2 Effects of Progressive Deformation

The effects of progressive deformation which we see in a volume of rock depend on the character the flow had at any instant of its deformation history, and is in fact the 'sum' of all those different flow patterns (Fig. 3.4). Progressive deformation by any sequence of flow types in a volume of rock leads to a change in shape of the volume itself, and to changes in the rock fabric. In an originally homogeneous volume of rock, spherical objects may deform passively to become ellipsoidal (Fig. 3.1), oblong rigid objects may rotate and various types of foliations and/or lineations[1] develop (Hobbs et al., 1976; Williams, 1977). Foliations and lineations are some of the most important keys to unravel the deformation history of a volume of rock. In an inhomogeneous volume of rock, layers or dykes may become folded or boudinaged, or just flattened and rotated into a new orientation.

An inhomogeneous fabric in a volume of rock, such as a folded foliation, a boudinaged layer, or a network of shear zones, indicates that the flow must have been different from place to place in that volume of rock for at least part of the deformation history. However, it is usually possible to find domains on some scale in which foliations and lineations are straight, and in which deformation must have been approximately homogeneous. In such domains, the coaxial or non-coaxial nature of the flow can sometimes be determined from the symmetry of relatively small fabric elements in the rock. Two examples are given below, others can be found in Section 4.4 and in the literature (Simpson & Schmid, 1983; Passchier, 1986a).

---

[1] We use 'lineation' to mean a linear structure that occurs repetitively in a rock, e.g. an array of elongate parallel pebbles, or lines of intersection of two foliations. 'Foliation' is used in the sense of a planar structure which occurs repetitively within a rock. A foliation trace on an outcrop surface is not a lineation since it occurs only in one plane (the outcrop surface) and not throughout the rock.

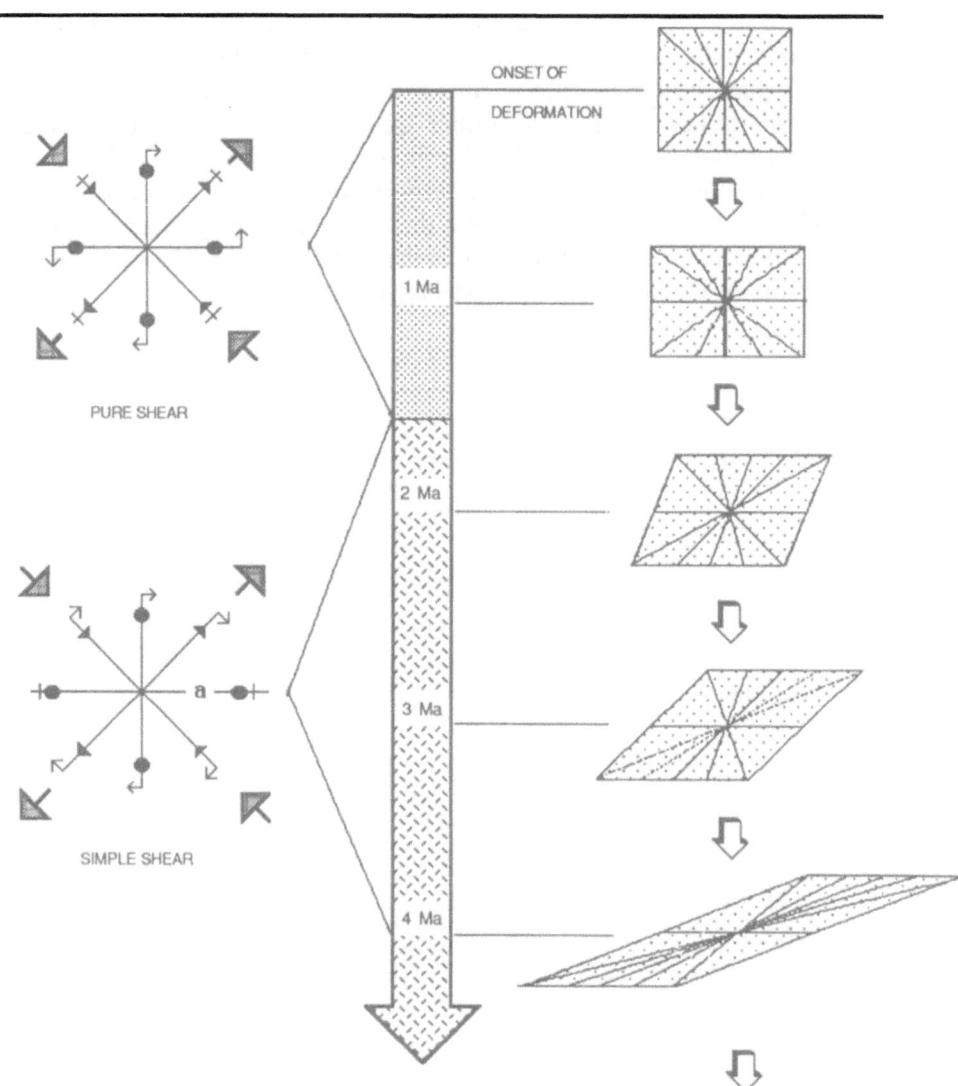

Fig. 3.4. How does 'flow' in the rock govern 'deformation'? Flow describes the velocity and direction of deforming lines at a specific instant in time. A deformation history can be thought of as a sequence of small deformation increments (e.g. lasting a few minutes on this time scale), during which a flow regime with constant parameters (such as vorticity, stretching rate) governs the change in shape of the rock. These flow parameters can be different for each subsequent deformation increment. The 'sum' of all the deformation increments results in the final structures observed by the geologist. In this figure an episode of pure shear flow accumulates deformation for 1.6 Ma, followed by 3 Ma of simple shear. PSA - principal stretching axes; a - lines which have zero-angular velocity.

20

(1) In simple shear, the distribution of small feldspar grains which form in the rim of a large feldspar porphyroclast will be 'asymmetrical' because the mass of small grains rotates away from the extensional principal stretching axis. In pure shear, it remains parallel to this axis (Fig. 3.5a).

(2) Tension gashes, infilled by either quartz or melt, tend to develop perpendicular to the extensional principal stretching axis (Fig. 3.5b; Etheridge, 1983). With progressive deformation, new sections of such veins develop in the same orientation, but the old sections may rotate if flow is persistently non-coaxial. The result is the familiar curved shape of tension gashes in shear zones (Ramsay & Huber, 1983).

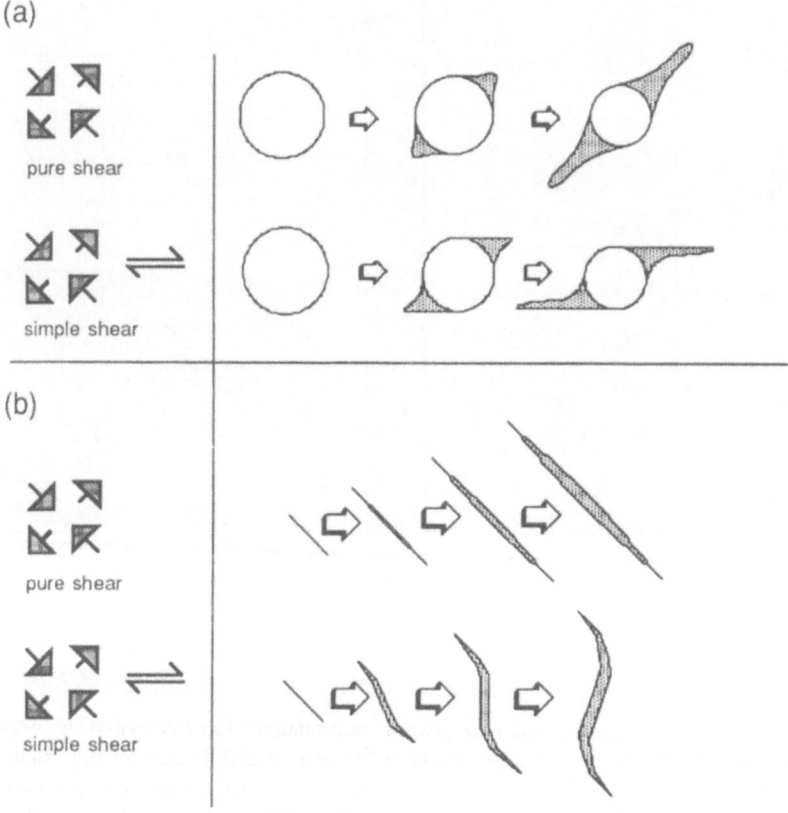

Fig. 3.5. Deformation paths accumulated by pure shear and simple shear in rocks can be distinguished by the symmetry of some final fabric elements; (a) Large phenocrysts in a fine grained matrix recrystallise at the edges and can develop symmetric or asymmetric 'tails' of recrystallised material; (b) tension gashes which open parallel to PSA of flow will remain symmetric during deformation by progressive pure shear, but will become asymmetric during simple shear due to the rotation of the central, first formed segment. Solid arrows at left indicate orientation of flow-PSA during progressive deformation.

3.3 Fabric Elements

3.3.1 Granoblastic Fabrics

Mineral grains in high-grade gneisses are usually bounded by straight or slightly bent grain boundaries which define a regular, polygonal pattern, known as a 'granoblastic' fabric (Fig. 3.6a). Internally, such grains are generally strain-free and show straight extinction under the microscope. This contrasts with fabrics in deformed low-grade metamorphic rocks which are generally made up of mineral grains with irregular grain boundaries and a deformed crystal lattice (Fig. 3.6b). In polymineralic rocks such as gneisses, the shape of individual mineral grains in a granoblastic fabric is defined by the mineral species in contact. 'Isotropic' minerals such as feldspars, quartz, carbonates, cordierite and garnets tend to form networks of equidimensional grains (Fig. 3.6a; white), pyroxenes and amphiboles are more oblong (Fig. 3.6a; grey), and 'anisotropic' minerals such as micas, sillimanite and tourmaline have a pronounced oblong (in many cases idiomorphic) shape (Fig. 3.6a; black), especially when isolated between quartz and feldspar grains.

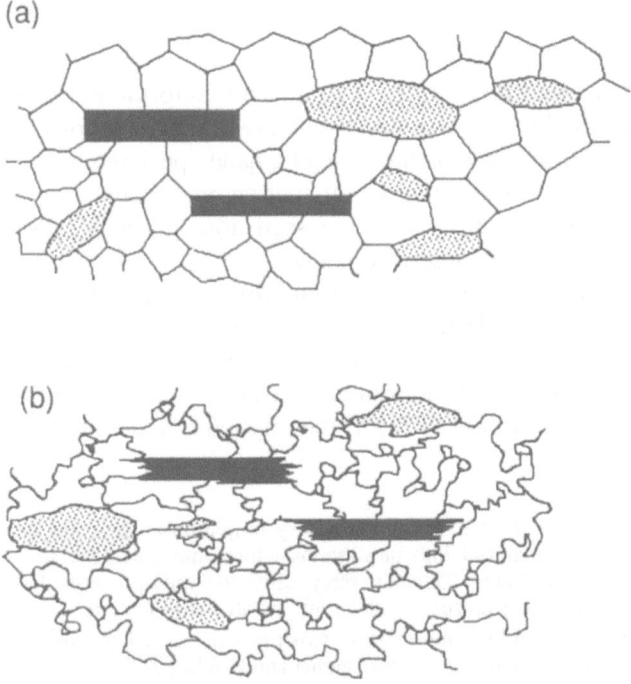

Fig. 3.6. (a) Typical microfabric of a high-grade rock with straight, short grain boundaries, also known as a granoblastic fabric. (b) Typical microfabric of a low-grade rock with serrate grain boundaries and numerous recrystallised grains. White - isotropic minerals (quartz, feldspars); grey - weakly anisotropic minerals (hornblende, pyroxenes); black - strongly anisotropic minerals (micas).

The granoblastic fabric described above develops in response to grain boundary migration towards a low-energy configuration driven by the free energy of the boundaries themselves; shorter (straight) boundaries have lower free energy than irregular ones (Vernon,1983; Poirier,1985). This process is counteracted by deformation which leads to accumulation of lattice defects and irregular grain boundaries. Granoblastic fabrics develop where recrystallisation[1] (Urai et al., 1986) and diffusion processes can proceed relatively fast, notably in high-grade metamorphic conditions. Development of melt pockets along grain boundaries may also play a role in the development of some granoblastic fabrics. Damage due to grain-scale deformation mechanisms is usually 'repaired' by grain boundary migration before a high-grade gneiss is uplifted to low-grade conditions. This explains why apparently undeformed granoblastic fabrics are common in gneisses which show evidence of strong macroscopic deformation, such as isoclinal folding.

The coarse grain-size in many high-grade rocks with granoblastic fabrics implies that few microscopic structural details will be visible in thin section, especially in quartzo-feldspathic rocks[2]. Structural relations should therefore be determined as much as possible in the field; there is little hope that any relations which have not been established in the field will be clear in thin section.

3.3.2 Shape and Mineral Fabrics

Fabric elements that result from crystalplastic deformation under high-grade metamorphic conditions can be divided into two groups: fabric elements that are homogeneously distributed on the scale of a hand-specimen, and fabric elements that are inhomogeneously distributed such as mesoscopic 'augen' structures, folds and boudins. The most commonly seen 'homogeneous' fabric elements are shown in Fig. 3.7 and are described below:
(1) A crystallographic preferred orientation of equidimensional minerals deformed by crystalplastic deformation (Fig. 3.7a);
(2) A preferred orientation of elongate or planar grains in response to rotation or new growth. This fabric element can define a lineation (mineral lineation) or foliation[3] in a rock (Fig. 3.7b). If the original orientation distribution of the

---

[1] Recrystallisation is a process of grain boundary migration in the solid state which causes rearrangement of the material into new grains which may have the same or a different composition from the old ones. 'Dynamic recrystallisation' only proceeds during deformation, while 'static recrystallisation' proceeds both during and after deformation.

[2] The best chance of finding microscopic structures preserved in high-grade rocks is in metapelites, in minerals such as cordierite, garnet and K-feldspar. Such rocks should therefore always be sampled.

[3] Such a foliation is usually named after the constituent mineral, e.g. a mica-foliation (Fig. 3.7b), a hornblende-foliation, etc. Notice that this is not the same as a foliation defined by elongate aggregates of the same minerals (a planar shape fabric); in that case, individual mica or hornblende grains may have various or even random orientations within the mica or hornblende aggregates.

grains was random, a foliation follows the XY plane, and a lineation the X direction of the strain ellipsoid (with $X \geq Y \geq Z$). Mineral lineations are usually defined by elongate grains such as hornblende and tourmaline, but may also be defined by planar crystals such as biotite, provided the crystals share a common axis in their preferred orientation.

(3) A shape fabric of linear or planar crystal aggregates that can form a lineation (linear shape fabric or stretching lineation[1]) and foliation (planar shape

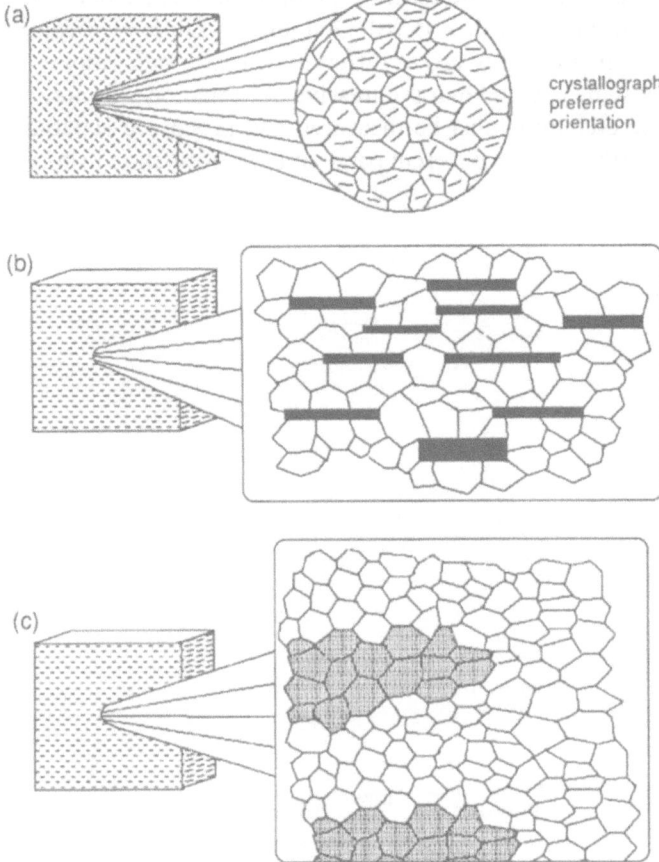

Fig. 3.7a-c. Examples of homogeneous fabric elements that can be found in high-grade gneisses; (a) crystallographic preferred orientation (indicated by the black lines in individual grains) in equidimensional grains with polygonal, straight grain boundaries; (b) crystallographic preferred orientation of tabular (black) grains in equidimensional (white), polygonal grains. Both kinds of crystals have straight grain boundaries; (c) two phases of equidimensional grains. A shape fabric is defined by elongate clusters of dark equidimensional grains. Both phases have polygonal, straight grain boundaries.

---

[1] The word 'linear shape fabric' is a general term which covers any type of elongate linear aggregate, including stretching lineations and some intersection lineations formed by disruption of layering. The word 'stretching lineation' is reserved for a lineation of aggregates which formed by stretching of equidimensional or elongate objects.

fabric) in a rock (Fig. 3.7c). If the shape fabric develops from originally equidimensional features, these lineations or foliations follow the orientation of the X direction or the XY plane of the strain ellipsoid, respectively.

(4) A compositional layering formed by flattening of primary layering or intrusive veins, and layering formed by metamorphic differentiation (Fig. 3.7d).

(5) A fabric defined by elongated pockets of partial melt which formed in situ.

In any particular gneiss, the fabric elements mentioned above can coexist and are not necessarily parallel (Fig. 3.7d). Obliquity of fabric elements of different kinds can mean that (1) they are of different or partly overlapping ages, or (2) they are of the same age and that progressive deformation was non-coaxial. The different fabric elements must be measured and processed separately, as they may independently relate to different parts of the deformation history of a rock.

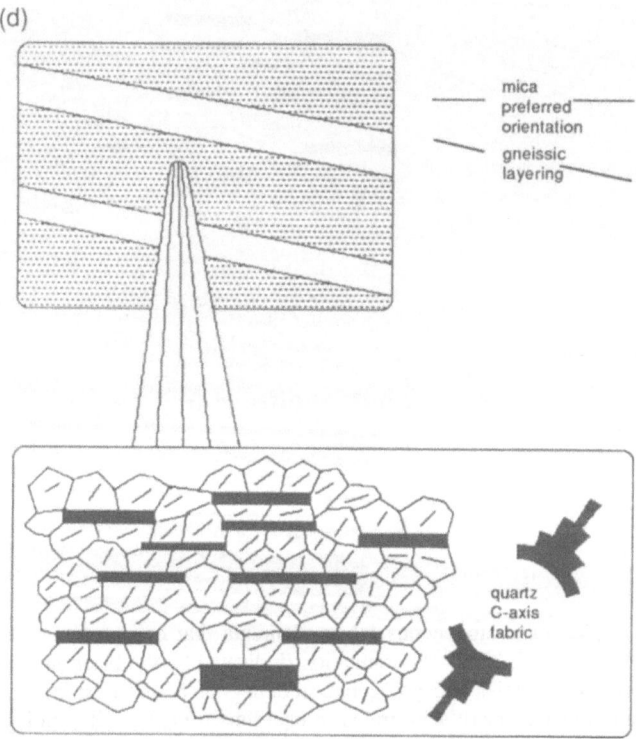

Fig. 3.7d. Three coexisting fabric elements with different orientations in the same rock: mica-foliation, gneissose layering and quartz C-axis preferred orientation.

3.3.3 Layering[1]

Layering in gneisses can be a primary structure of sedimentary or igneous origin, or it may be of secondary origin and the result of either intense deformation or solid state metamorphic differentiation. In many cases it is a combination of these features.

Layering in Metasedimentary Rocks

Survival of primary sedimentary features depends on the intensity of deformation and recrystallisation versus the scale of the primary features. Primary bedding can survive high amounts of deformation and high-grade metamorphism, but the primary stratigraphic sequence often becomes very much disturbed. During deformation, bedding is generally accentuated by one or more of the following processes: attenuation leading to thinner but more pronounced beds; the development of a bedding-parallel foliation; the development of bedding-parallel segregations (generally quartzo-feldspathic pegmatite with or without biotite-rich selvages or, in some instances, concentrations of various other minerals or mineral assemblages); the intrusion of bedding-parallel veins of pegmatite or other igneous rocks.

Psammitic metasedimentary rocks may develop quartz or melt veins in various structural positions such as tension gashes and boudin necks. With increasing deformation these veins become subparallel and streaked out into thin discontinuous, subparallel layers. Pelitic metasedimentary rocks generally develop quartzo-feldspathic melt veins, again in a variety of primary orientations that become subparallel as deformation increases.

Layering in Igneous Rocks

Some undeformed intrusive igneous rocks display regular layering which formed during crystallisation of the magma. Such layering can be a rhythmic alternation of minerals of different size and/or composition and may even be graded. In some cases it bears a striking resemblance to sedimentary layering; even features such as cross-bedding, convoluted layering and loadcasts occur in igneous rocks. Explanations of how these structures form, as given in Parsons (1987) and Paterson (1988), are:
(1) 'Layer-by-layer' deposition due to gravity-controlled crystal settling and variations in crystal supply in a cooling magma, possibly aided by density-induced liquid fractionation; deposition by a sequence of magmatic density

---

[1] Layering in gneisses is also referred to as 'banding', as structures in gneissic rocks are often described from planar outcrop surfaces. However, banding in gneisses is really the two-dimensional expression of layering, which is a three-dimensional structure.

currents; or periodic nucleation, resorption and coarsening near the solidus front.

(2) In-situ solidification of a magma sequence that has become compositionally stratified.

(3) Sequential injection of several planar veins or sheets of magma along the same channel or parallel channels.

(4) Preferential new growth of minerals during metamorphism, which can . strengthen existing patterns of layering in undeformed igneous rocks.

Layering as an Effect of Deformation

Gneissose layering can reflect primary features, but may also develop by extreme flattening (ductile deformation) of compositional inhomogeneities in the original rock, which were not necessarily planar. The intensity of deformation in many gneisses may not be immediately obvious to those who are not familiar with such rocks (Figs. 1.1; 3.8). It is not widely appreciated that igneous rocks can easily be converted into well-layered gneisses by intense deformation, and so we demonstrate this process by a series of illustrations. The following principal situations can be distinguished:

(1) Homogeneous deformation of vein networks such as pegmatite veins or amphibolite dykes in a relatively massive host rock such as granite. This leads to the formation of pegmatite-layered or amphibolite-layered gneiss (Figs. 3.8a and 3.9).

(2) Homogeneous deformation of rock fragments such as amphibolite xenoliths in granite. This can produce a strongly layered rock (Fig. 3.8b).

(3) Homogeneous deformation of a homogeneous igneous rock with large phenocrysts (Fig. 3.8c). Examples are shown of a porphyritic granite (Fig. 3.10 a-e) and an anorthosite with leucogabbroic patches (Fig. 3.11 a-d).

(4) Inhomogeneous deformation of a homogeneous rock (e.g. gabbro) (Fig. 3.8d). Lenses or layers of the original rock may remain as remnants between strongly deformed zones with fairly sharp boundaries. These lenses or layers could be mistaken as younger, undeformed intrusions into an older gneiss. Figure 3.12 illustrates such inhomogeneous deformation of an originally homogeneous leucogabbro.

It is usually impossible to determine the origin of a gneiss from the intensely deformed end-product of these processes in the field. For example, a similar layering to that shown in Fig. 3.11d can also be produced by: intense deformation of a mafic volcanic rock (Fig. 3.13); shearing, isoclinal folding and flattening of a contact between igneous or gneissic basement rocks and overlying metamorphosed sedimentary rocks (Fig. 3.14); and the rotation and distortion of amphibolite dykes or dykelets in a granitic rock by the stages described in (1) above. During rotation and distortion, some dykes would probably be boudinaged and could resemble the situation in Fig. 3.8b3.

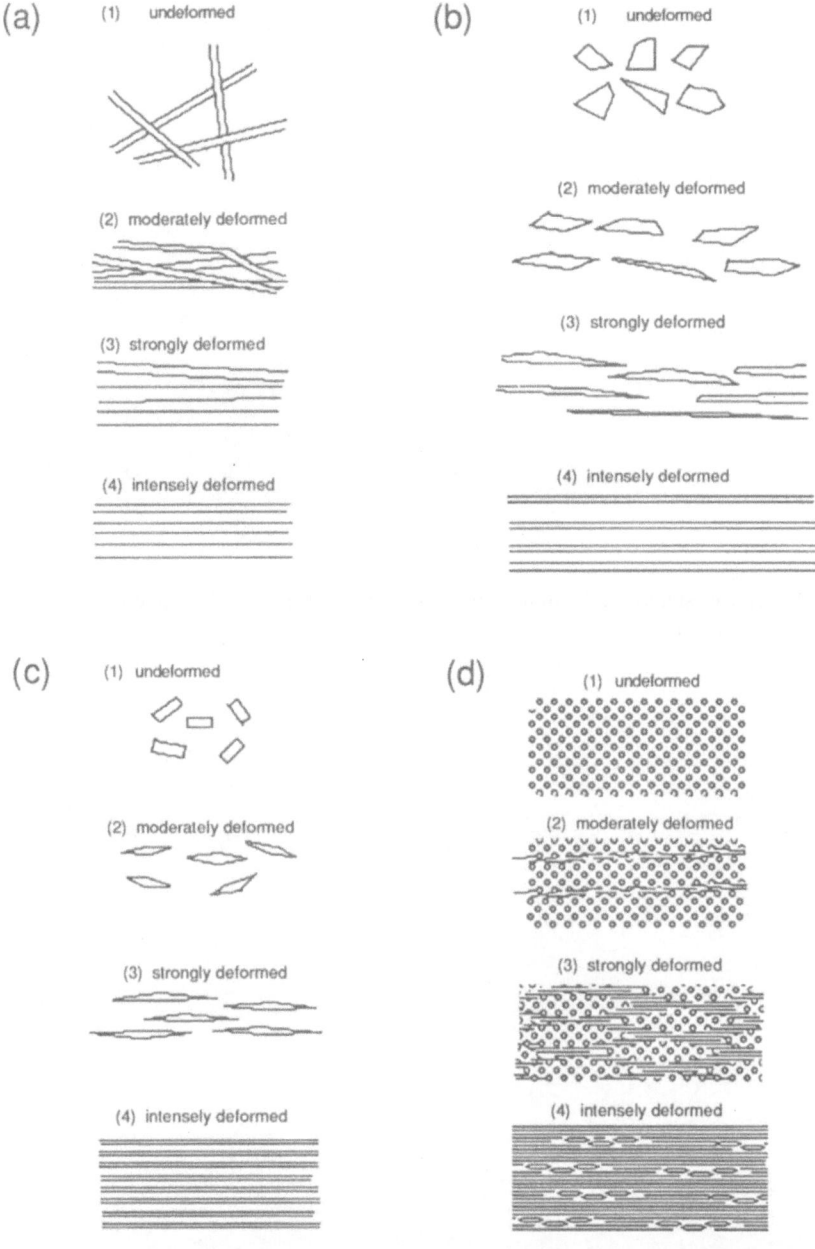

Fig. 3.8. Four examples of outcrop-scale progressive deformation typical of high-grade gneiss terrains, all leading to a uniform, parallel-layered gneiss; (a) homogeneous deformation of vein networks; (b) homogeneous deformation of rock fragments; (c) homogeneous deformation of a homogeneous igneous rock with heterogeneous grain size, e.g. porphyritic granite; (d) inhomogeneous deformation of a homogeneous igneous rock, e.g. gabbro.

28

Fig. 3.9. Typical, superficially simple, pegmatite-layered quartzo-feldspathic gneiss. Fisken-æsset, SW Greenland.

Figs. 3.10a-e. Sequence of outcrops of deformed porphyritic granite representing a strain gradient from undeformed granite to layered gneiss. Gotthard Massif, Switzerland. Scale bar in cm. Photographs by J.D. Hock, all same scale. **(a)** Undeformed granite.

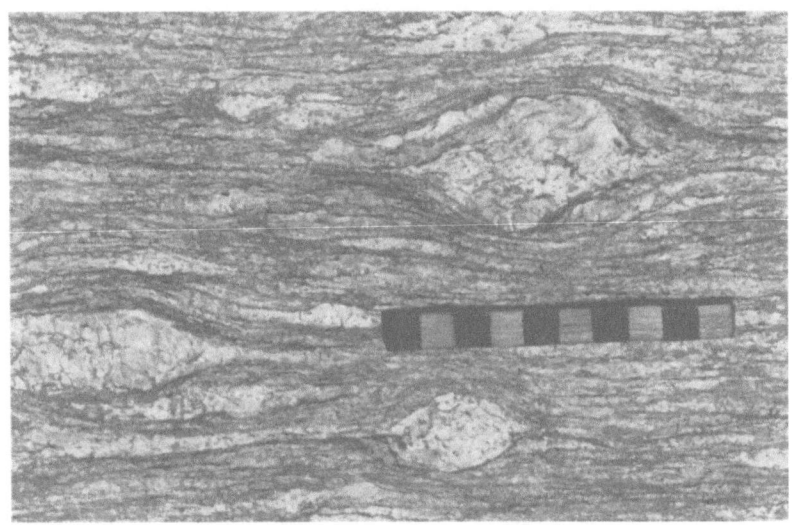

Fig. 3.10b. Augen gneiss.

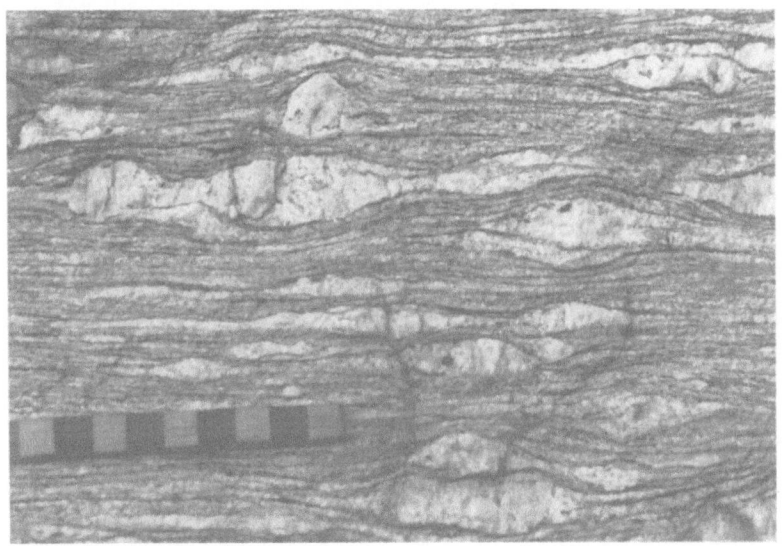

Fig. 3.10c. Augen gneiss with flattened K-feldspar porphyroclasts.

Fig. 3.10d. Layered gneiss with still recognisable strongly flattened K-feldspar porphyroclasts.

Fig. 3.10e. Finely layered gneiss.

Figs. 3.11a-d. Sequence of outcrops of anorthosite with relic pyroxene oikocrysts, showing a gradation from undeformed anorthosite with increasing deformation to tectonically layered, intensely deformed anorthosite. All photographs at same scale. Fiskenæsset, SW Greenland. (a) Undeformed anorthosite.

Fig. 3.11b. Weakly deformed anorthosite with streaked-out pyroxene oikocrysts.

32

Fig. 3.11c. Strongly deformed oikocrystic anorthosite.

Fig. 3.11d. Tectonically layered, intensely deformed anorthosite with post-tectonic granoblastic recrystallisation.

Fig. 3.12. Inhomogeneously deformed leucogabbro; undeformed lens with plagioclase mega-crysts in strongly foliated and tectonically layered leucogabbro that was originally like the undeformed portion. Casual observation might lead to the erroneous interpretation that the undeformed leucogabbro intruded an older gneiss. Fiskenæsset, SW Greenland.

Fig. 3.13. Layered amphibolite derived by strong deformation of pillow lava. Barö, Åland, SW Finland.

Fig. 3.14. Example of the superficially simple end-product of intense, complex deformation of a gneissic basement (dark Lewisian orthogneiss) and metasedimentary cover (light Moine psammite). The rocks have been deformed into thin, attenuated parallel layers and then isoclinally folded. Borgie, N-Scotland.

Other Mechanisms that Form Layering

In some rocks, layering overprints older igneous or sedimentary structures in such a way that it cannot be explained by simple flattening of fabric elements. Some micaceous quartzo-feldspathic and pelitic rocks start to melt in high-grade metamorphic conditions. The melt may accumulate in intragranular melt-pockets while most of the rock is still in a solid state. Such a melt fraction obscures older fabric elements in a rock, but need not destroy them. It can produce a vague light-coloured layering or sets of leucocratic veins at an angle to older fabric elements (Figs. 4.25; 8.7). In many cases, such veins or layers are parallel to the axial plane of folds (Hudleston, 1989).

A similar vague layering occurs in some igneous rocks as an effect of late-stage differentiation in a cooling and solidifying magma; crystals grow preferentially at some levels in reaction with the remaining liquid. Reactions among minerals and between minerals and infiltrating fluids in the sub-solidus state may produce or intensify layering.

Another mechanism for producing gneissose layering was suggested by van der Molen (1985), who claimed that some gneissose layering formed by differentiation at medium- to high-grade metamorphic conditions. The process involves the solution and redeposition of material, driven by the local stress field. In this model, a homogeneous rock can be transformed into a layered rock without the formation of a melt phase and without strong deformation.

3.3.4 Augen Structures in Gneiss

Many gneisses contain lens-shaped single crystals or coarse crystal aggregates in a finer-grained matrix. Such structures are called 'augen', and a gneiss with abundant 'augen' (generally feldspar) is termed an 'augen gneiss' (Fig. 3.10b). There has been considerable debate as to whether augen in gneiss develop during or after deformation, or reflect modified primary crystals. From numerous recent field and microstructural studies it seems that the latter case is most common. In many feldspar augen, a core with original igneous zoning is truncated by a lensoid outer metamorphic rim. A zoning of metamorphic origin parallel to the rim of the augen has so far only been observed in small feldspar augen of some mylonites (Wintsch & Knipe, 1983). Most augen gneiss probably developed from coarse porphyritic granitoid rocks in which the grain size was gradually reduced during dynamic recrystallisation of quartz and feldspar (Fig 3.10a-e; Tullis et al., 1982). Deformation and recrystallisation mainly affected the margins of large feldspar phenocrysts which were gradually reduced in size and modified from an idiomorphic to a lensoid shape. Locally, extreme boudinage and recrystallisation of coarse-grained veins such as pegmatites also lead to the development of augen.

3.4 Shear Zones in High-Grade Conditions

3.4.1 Introduction

Shear zones are relatively narrow planar zones of high ductile strain between less deformed wall rocks, across which markers, such as layers or veins, are displaced. They are the ductile analogue of brittle fault zones, and occur in many high-grade gneiss terrains. Shear zones generally contain a foliation subparallel to the principal plane of the strain ellipsoid and perpendicular to the direction of maximum shortening (Fig. 3.15).

Many high-grade gneiss terrains show evidence that flow in shear zones was non-coaxial; such evidence includes deflection of newly formed foliation at the edge of the zone (Fig. 3.16) and the presence of fabric elements with monoclinic shape symmetry[1] (Figs. 3.17; 4.11).

---

[1] We mean the symmetry of the fabric itself, not crystallographic symmetry. The monoclinic shape symmetry of fabrics is the result of progressive non-coaxial deformation.

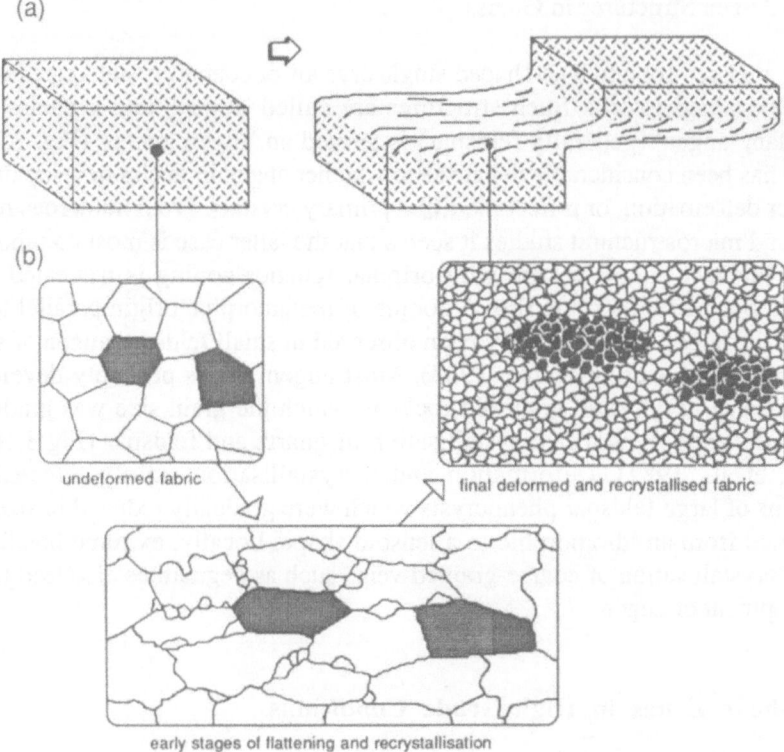

Fig.3.15. (a) Ductile shear zone between rigid wall rocks; (b) Stages of crystalplastic deformation and recrystallisation in a shear zone. A coarse grained equigranular aggregate of dark and light minerals is converted into a fine grained aggregate with a shape fabric.

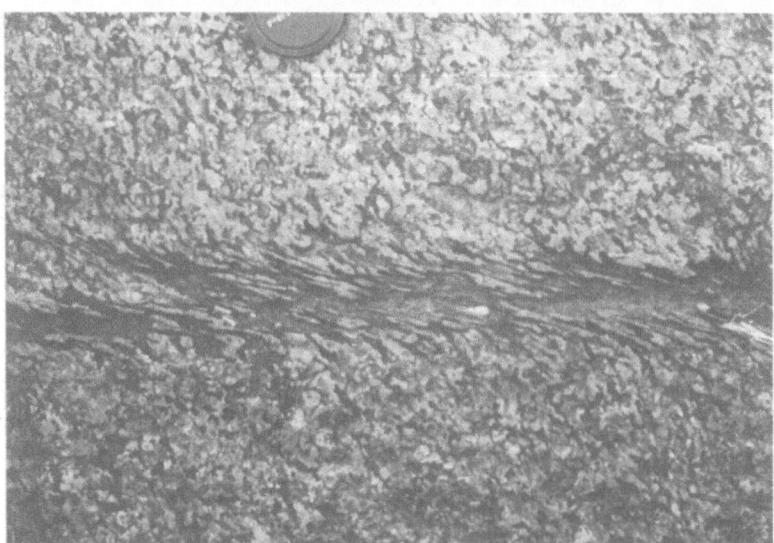

Fig. 3.16. Ductile shear zone in a granite. Duchess, Mt. Isa Inlier, Australia. Sinistral shear sense.

3.4.2 Development of Fabrics in Shear Zones

In shear zones, single crystals or crystal aggregates are flattened and stretched into lenses and aggregates which define a shape fabric (Fig. 3.18). A crystallographic preferred orientation of quartz, micas and, to a lesser extent, feldspars is usually developed. The contact of ductile shear zones and wall rock is a gradual fabric transition. Originally planar or linear structures will become extended or shortened, depending on their initial orientation (Figs. 3.20; 4.28; 4.29); this leads to folding or boudinage if a competency contrast exists (Fig. 3.20). However, at high strain values in shear zones, all veins tend towards parallelism, and former boudins or isoclinal folds can become so strongly flattened that their original shape is lost; they may appear as an undeformed small-scale planar layering (Fig. 3.20). A biotite-, hornblende- or pyroxene-foliation will also tend towards parallelism with strongly flattened layering in high-strain shear zones. Highly deformed rocks from ductile shear zones are known as *mylonites*[1] (Bell & Etheridge, 1973; Hobbs et al, 1976; White et al, 1980; Tullis et al 1982).

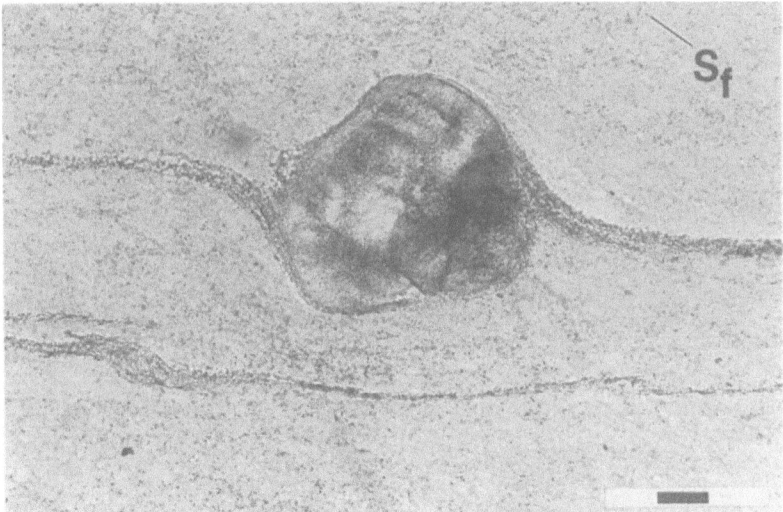

Fig. 3.17. Micrograph showing asymmetric tails of recrystallised feldspar around a feldspar porphyroclast in quartzite. Note the 'stair stepping' of the tails and the oblique fabric in the quartzite. Sinistral shear sense. Sf - foliation defined by elongated small quartz grains. Scale bar 0.1 mm. St Barthélemy Gneiss, Tarascon, France.

---

[1] Although the word mylonite stems from the Greek word 'mylon', a mill, there is not necessarily any brittle deformation associated with it. At low- to medium-grade metamorphism, mylonites generally have a smaller grain size than the less deformed wall rock; however, in high-grade environments this is not necessarily the case because deformation is generally followed by static recrystallisation.

38

Fig. 3.18. Stretching lineation in a ductile shear zone in granite. St Barthélemy Gneiss, Tarascon, France.

Fig. 3.19. Deformed leucogabbro with relict 10 cm diameter cumulus plagioclase crystals forming oblate ellipsoids. Fiskenæsset, SW Greenland.

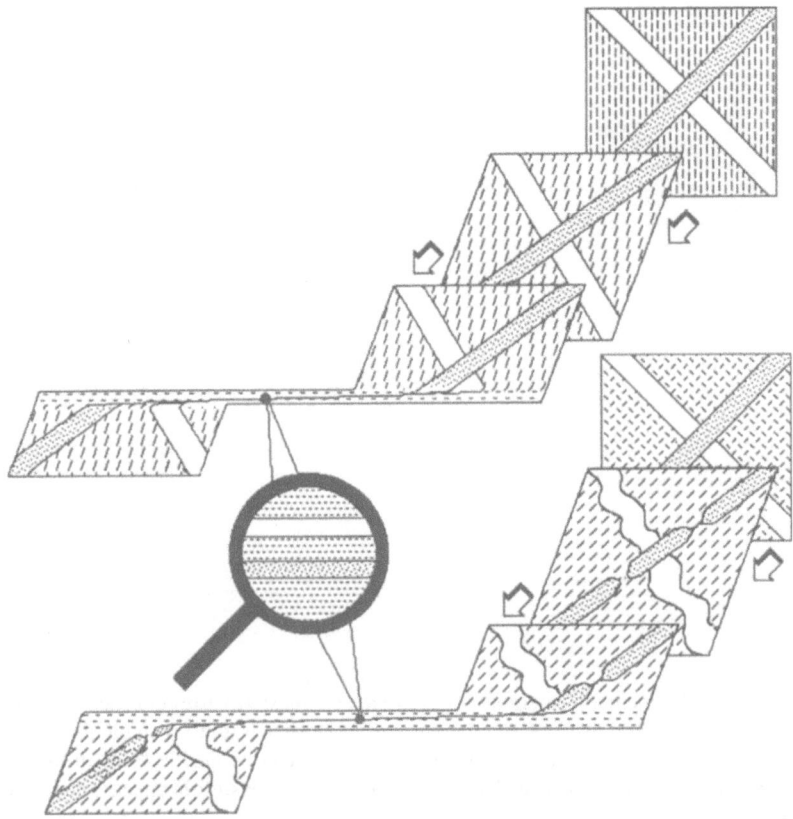

Fig. 3.20. Early stages of a deformation history may be preserved in the little strained wall rocks of a shear zone, but are usually destroyed by high strain within a shear zone where all fabric elements are rotated towards parallelism and strongly flattened. This common end-product is shown above developing from two rocks with different initial fabrics and deformation histories.

3.4.3 Isoclinal Folds in Shear Zones

Isoclinal folds with complex three-dimensional shape develop in shear zones in a number of ways. Tubular or sheath folds form by flattening and attenuation of pre-existing folds, or by rotation of layer segments in response to the presence of a strain-rate gradient oblique to foliation planes (Cobbold & Quinquis, 1979; Platt, 1983; Fig. 3.21a). Buckling due to minor transient shortening in the foliation plane oblique to the main movement direction, i.e. deviation from plane strain flow (Fig. 3.21b), can produce isoclinal folds known as oblique folds (Passchier, 1986a). All these isoclinal fold types have a dominant orientation of limb segments parallel to the stretching lineation.

(a) sheath folds

(simple shear)

(b) oblique folds

(simple shear + constriction)

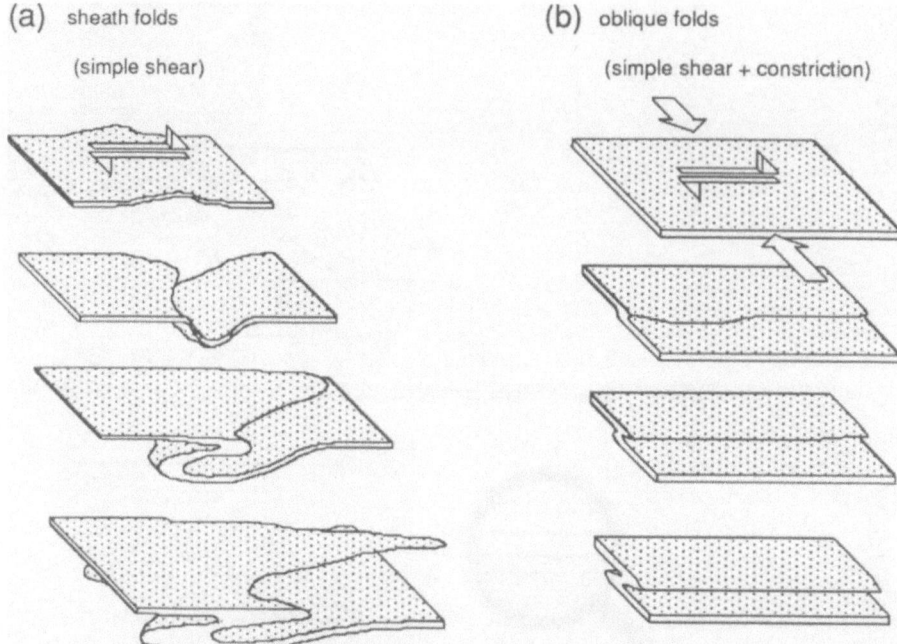

Fig. 3.21. Development of isoclinal folds in shear zones; (a) irregularities in a foliation plane parallel to the flow plane of simple shear result in 'sheath folds' with tubular shape; (b) a component of shortening oblique to the shear direction causes buckle folds in the foliation which can stretch to become isoclinal 'oblique folds'. Both types of fold appear similar in cross-section.

3.5 Fabric Distribution in Shear Zones

3.5.1 Introduction

In shear zones, the active deformation mechanisms depend on the rock composition, the local temperature and lithostatic pressure, composition and pressure of the metamorphic fluid, and on the bulk imposed strain rate (Fig. 3.22). The fabrics formed in shear zones will therefore also depend on these parameters. In the previous section, we discussed mylonite fabrics that developed under medium to high-grade metamorphic conditions; in this section we describe deformation regimes and fabrics that developed under lower-grade conditions at various depths in the crust.

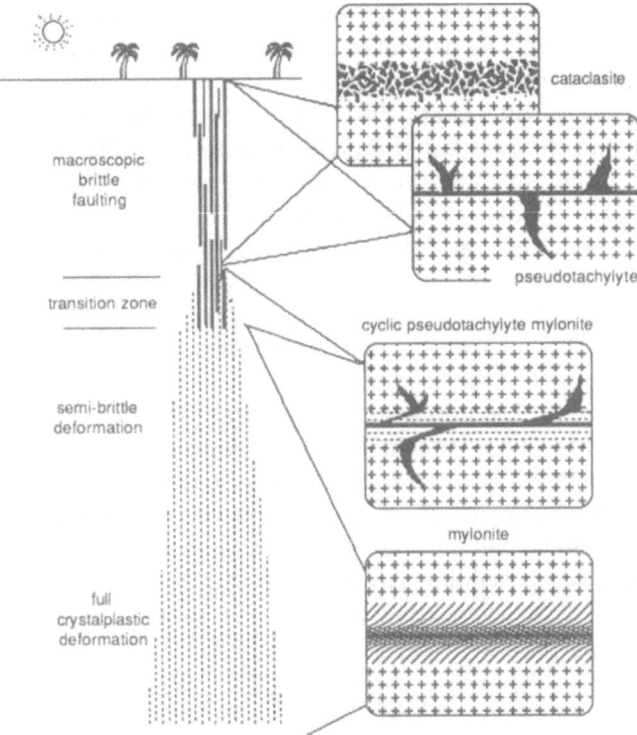

Fig. 3.22. Distribution of different kinds of 'fault rocks' with depth in the crust along a major transcurrent shear zone. Cataclasite and pseudotachylyte form at high levels, cyclic pseudotachylyte-mylonite at a deeper level and various types of mylonite at the deepest level. Deformation regimes are given on the left.

3.5.2 The Semi-Brittle Deformation Regime

Although many parameters influence the deformation mechanisms in shear zones, the following trend can be observed in many gneiss terrains. During high-grade metamorphic conditions, shear zones in gneisses tend to be relatively wide, and the rock-forming minerals such as quartz, feldspars, micas and hornblende deform dominantly by crystalplastic flow (Behrmann & Mainprice, 1987). Flow tends to be relatively homogeneous on a small scale, and mylonites develop. Under medium- to low-grade metamorphic conditions, shear zones tend to be of more restricted width[1], and although the contact between shear zones and wall rocks is a gradual fabric transition, it is more sharply defined than in high-grade ductile shear zones. Flow in these zones can be macroscopically homogeneous and cohesive as at high metamorphic grade, but may be

---

[1] Relatively wide' and 'relatively narrow' are used in the sense of width/finite strain ratio; for a certain amount of displacement along a shear zone (and finite strain attained in the zone), shear zones tend to be wider in high-grade conditions than in low-grade conditions. Faults lie at the ultimate end of this trend.

inhomogeneous on a small scale (Kerrich et al., 1980); some mineral phases such as quartz and micas will deform by crystalplastic flow while others such as feldspar and hornblende deform predominantly by microfracturing (cf. Sibson, 1977; Grocott, 1977; Tullis et al., 1982). This deformation regime is therefore known as the *semi-brittle regime* (Fig. 3.22; Hobbs et al., 1986; Strehlau, 1986; Scholz, 1988). In polymineralic rocks there is obviously a gradual transition between the domain of full crystalline plasticity and this semi-brittle regime.

Mylonites which develop in the semi-brittle regime have a characteristic fabric with two main components: porphyroclasts[1] and a matrix. The matrix or groundmass comprises a finely layered and fine-grained, dynamically recrystallised material with a planar and linear shape fabric. The matrix wraps around rounded porphyroclasts of the 'harder phases' that survive as non-recrystallised relics of the host rock. The porphyroclasts generally have a monoclinic shape symmetry (Figs. 3.17; 3.23; Passchier, 1986a; White et al., 1980; Tullis et al., 1982). A crystallographic preferred orientation of quartz, mica and, to a lesser extent, recrystallised feldspar is usually well developed in the groundmass. Some of the best known and most reliable sense of shear markers developed in shear zones under these conditions, as summarised in Section 4.4.

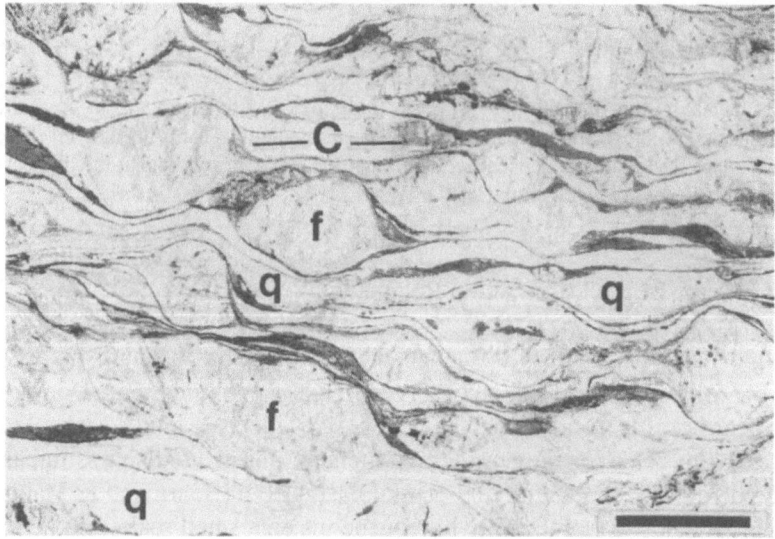

Fig. 3.23. Typical microstructure of a mylonite from the semi-brittle regime; a regular layering of partly recrystallised, flattened old grains of quartz (q) and biotite (dark grey) bend around rigid, brittly deforming feldspar porphyroclasts (f). In the centre of the photograph a shear band (c) transects the main foliation, indicating sinistral sense of shear. St. Barthélemy Gneiss, Tarascon, France.

---

[1] Porphyroclasts are grains that are significantly larger than those of the matrix and which are thought to represent relatively rigid relics of large grains in a deformed matrix; porphyroblasts are grains that are significantly larger than those of the matrix and which are thought to have grown in the solid state.

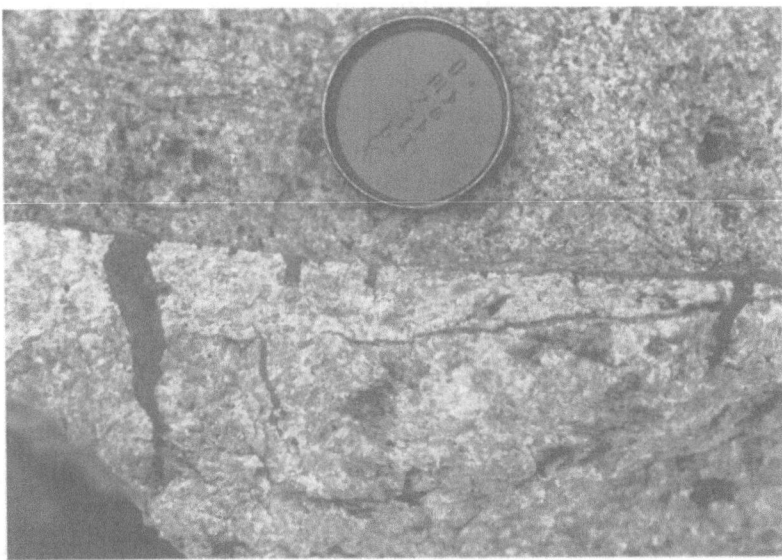

Fig. 3.24. Pseudotachylyte-generation surface separating gneisses with different composition. Injection veins extend from the generation surface into the wall rock. Vestfold Hills, Antarctica.

3.5.3 Cataclasite and Pseudotachylyte

Under very low-grade metamorphic conditions, deformation dominantly occurs by macroscopic brittle fracturing. This is the domain of unstable, seismogenic stick-slip deformation (Sibson, 1982; Meissner & Strehlau, 1982; Scholz, 1988), i.e. movement on discrete fault surfaces at seismic strain rates of several mm to m per second, alternating with long quiet periods of slow stress build-up. Ductile and semi-brittle shear zones are thought to deform at a steady strain rate of several mm to cm per year by ductile deformation processes without major breaks in the coherence of the rock. In fact, mylonites and brittle fault rocks can be formed along the same large scale shear zone if such a zone extends through a considerable section of the crust (Fig. 3.22).

Two basic types of fault rocks can be formed within the regime of discrete brittle faulting: cataclasite and pseudotachylyte. Cataclasite develops by brecciation of intact rock, generally with abundant fluid access, leading to breccias with abundant quartz veining. Pseudotachylyte forms by local melting of the rock along a brittle fault plane due to heat generated by frictional sliding (Phillpotts 1964; Francis & Sibson, 1973; Sibson 1974, 1975, 1977; Grocott 1981; Maddock, 1983, 1986; Maddock et al.,1987) or, possibly, in some cases by intense cataclasis (Wenk, 1978). Melting is thought to occur at temperatures exceeding $1000^{\circ}C$ in a zone a few mm wide called the 'generation surface'. Some of the melt may intrude minor branching faults which splay from the generation surface, and which are called 'injection veins'. The small volume of melt formed in this process cools rapidly to the temperature of the host rock. As a result, the melt quenches to a glass or very fine-grained material which occurs along fault planes and adjacent branching injection veins (Fig. 3.24). The very fine-grained or glassy groundmass may contain isolated fragments of mostly quartz and feldspar and, in relatively few cases, Fe-Mg rich silicates. The contacts between the groundmass and the wall rock are very sharp, even in thin section. Pseudotachylyte is rarely seen to be associated with quartz veins and generally occurs in massive, dry, low-porosity rocks such as gabbro, gneiss and amphibolite. This is because the fluid present in porous rocks lowers the effective normal stress over a fault plane upon heating and, consequently, not enough frictional heat can be produced to cause local melting. Therefore, pseudotachylyte never forms in porous sedimentary rocks.

3.5.4 The Transition Zone

In quartzo-feldspathic rocks, the transition between the domain of full crystalline plasticity and the semi-brittle regime is generally gradual. In contrast, the transition between the semi-brittle regime and the seismogenic, fully brittle domain where breccia and pseudotachylyte are formed is relatively sharp (Fig. 3.22; Sibson, 1982; Meissner & Strehlau, 1982; Tullis & Yund, 1987; Scholz, 1988). Nevertheless, it should not be imagined as a sharply defined plane in the crust; the local depth of the transition depends on the geothermal gradient, the rock composition, the orientation of compositional layering and other fabric elements, grain size, the interstitial fluid composition and pressure, bulk strain rate, and the orientation of the local stress field (Sibson, 1982; 1983; see below). Within any segment of the lithosphere, and even within a steeply dipping large shear zone, the transition will therefore occur in a zone of complex geometry.

The transition zone seems to be a domain where mylonite generation alternates with seismic faulting (Sibson, 1980; Passchier, 1982; 1984; 1986b). Shear zones have been described from the Outer Hebrides (Sibson, 1980) and central Australia (Allen, 1979) in which several faulting events were followed by ductile deformation in newly formed brittle fault rock. This apparently occurred under constant P-T conditions, and as part of a single continuous phase of deformation. Pseudotachylyte that underwent such ductile reactivation has been described from several areas (Watts & Williams,1979; Passchier, 1982; 1984). The alternating sequence of brittle and ductile events seems to result from the fine-grained nature of the pseudotachylyte matrix. In such a matrix, ductile deformation is possible under metamorphic conditions and at differential stress levels which are insufficient to create ductile flow in the more coarse-grained wall rocks (Passchier, 1982).

Hobbs & Ord (1988) have suggested that ductile reactivation of pseudo-tachylyte may not be restricted to the transition zone between the semi-brittle and seismogenic domains, but also occurs at greater depth within the semi-brittle domain itself due to transient unstable brittle collapse of cohesive steadily flowing shear zones. This alternative mechanism for cyclic pseudotachylyte-mylonite formation equally implies that brittle and ductile events occur alternately as part of a single deformation phase at a specific level in the crust (Section 4.6.1).

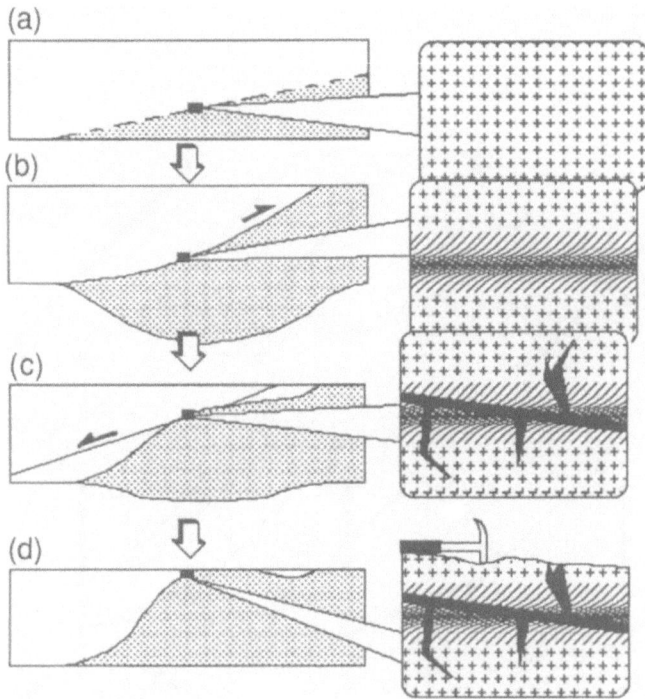

Fig. 3.25. Possible evolution of a shear zone in the continental crust. (a) A segment of the crust is transected by a ductile shear zone in which mylonite (b) formed; then (c) reactivation of this zone during uplift leads to pseudotachylyte transecting the older ductile mylonite, which is eventually (d) exposed at the surface.

3.5.5 Relation of Brittle and Ductile Fault Rocks in Gneiss Terrains

Many major shear zones in high-grade gneiss terrains appear to have a long record of activity. If gradual uplift and erosion accompany this deformation, the result will be that low-grade and macroscopically brittle (pseudotachylyte or cataclasite) structures will cut older ductile (mylonite) fabrics within a shear zone (Grocott, 1977; White et al., 1986; Fig. 3.25). In fact, many late cataclasite and pseudotachylyte veins in high-grade gneiss terrains are found within or adjacent to high- or medium-grade mylonite zones, often with similar displacement directions to the older mylonites. Their presence is thought to represent late stages of deformation during or after uplift of these high-grade terrains to shallow crustal levels. Except in the case of strike-slip zones, the distribution of brittle fault rocks and mylonites within outcrops of major shear zones is generally asymmetric (Fig. 3.26). This is due to the passive uplift of mylonites from deeper levels by continued movement along these zones (Fig. 3.25). A mylonite-dominated footwall results from normal movement on a shear zone; a mylonite-dominated hanging wall results from reverse movement. This relationship should be in agreement with the sense of shear determined from markers (Fig. 4.11), otherwise the shear zone has a more complex movement history.

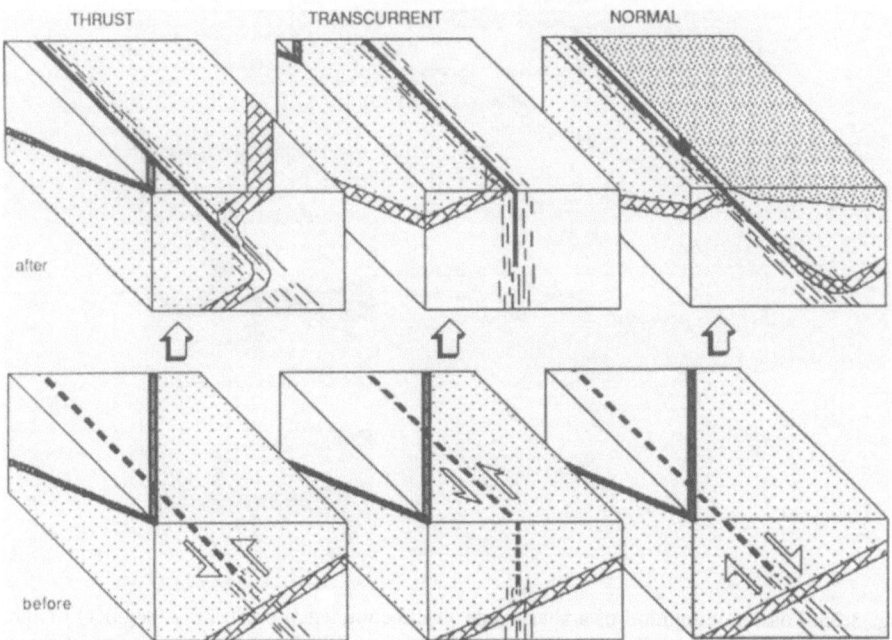

Fig. 3.26. Major thrust (reverse), transcurrent and normal (extensional) fault systems which cut a stratigraphic sequence (ornamented). The fault systems all consist of an upper brittle fault zone (solid line) which grades into a ductile shear zone at depth (solid dashed lines). Movement on the zones produces characteristic distribution patterns of brittle fault rocks (solid line) and uplifted, inactive mylonite (thin dashed lines); for transcurrent fault zones the distribution is symmetrical; in other cases the uplifted inactive mylonite lies predominantly in the footwall (normal fault) or hanging wall (thrust).

4 Interpretation of Structures and Fabrics
Recognition and Interpretation of Fabric Patterns in Outcrop

4.1 The Inadequate Memory of Rocks

High-grade gneisses with a long and complex history have a finite 'memory' of past events. This memory is formed by fabric elements such as foliations, lineations, folds, mineral assemblages, boudins and sequences of intrusion. One of the aims of a geologist is to tap this memory as effectively as possible. The memory of rocks, however, is rather inadequate in that it is partly destroyed by the same events which produce the fabric elements that are recorded (Williams, 1983). Strong deformation erases older fabric elements; intrusions, recrystal- lisation and partial melting do the same (Figs. 1.1; 4.1). One of the purposes of this manual is to help geologists working in high-grade gneiss terrains to recognise the effects which modify or obliterate the rock's memory, and to obtain all the data from the rock which are still available.

Strong deformation produces apparently simple fabrics by stretching and rotating existing structures into a subparallel orientation[1], and these could be misinterpreted. The best way to avoid such misinterpretation is by careful observation of structural relations in every outcrop, and by examination of a large area to see if a proposed sequence of events is everywhere confirmed. There are usually variations in strain magnitude over a large area, and structural geologists working in gneiss terrains should pay special attention to lenses or domains of low strain; such domains may preserve parts of the structural history which have been destroyed or obscured elsewhere (Figs. 4.1; 4.2).

In some cases, low-strain domains may be located by noting the width of certain rock units and the angle between veins and layers over a large area. If the width or angles increase in any direction then this may reflect decreasing strain. In many cases areas of low strain have an ellipsoid ('augen') shape, especially in major shear zones (Figs. 3.12; 4.1). Alternatively, they occur in protected sites such as inclusions in a major intrusion or adjacent to a relatively competent body such as a gabbro or anorthosite lens, and in the hinge zones of major folds. The structural relationships in such low strain domains should be carefully described

---

[1] This process is also known as 'transposition' (Hobbs et al., 1976; their Section 5.4).

48

and compared with those in other low strain areas of similar rocks. Eventually, their content should be compared with that of surrounding high strain areas to check if the differences can be explained by variations in finite strain.

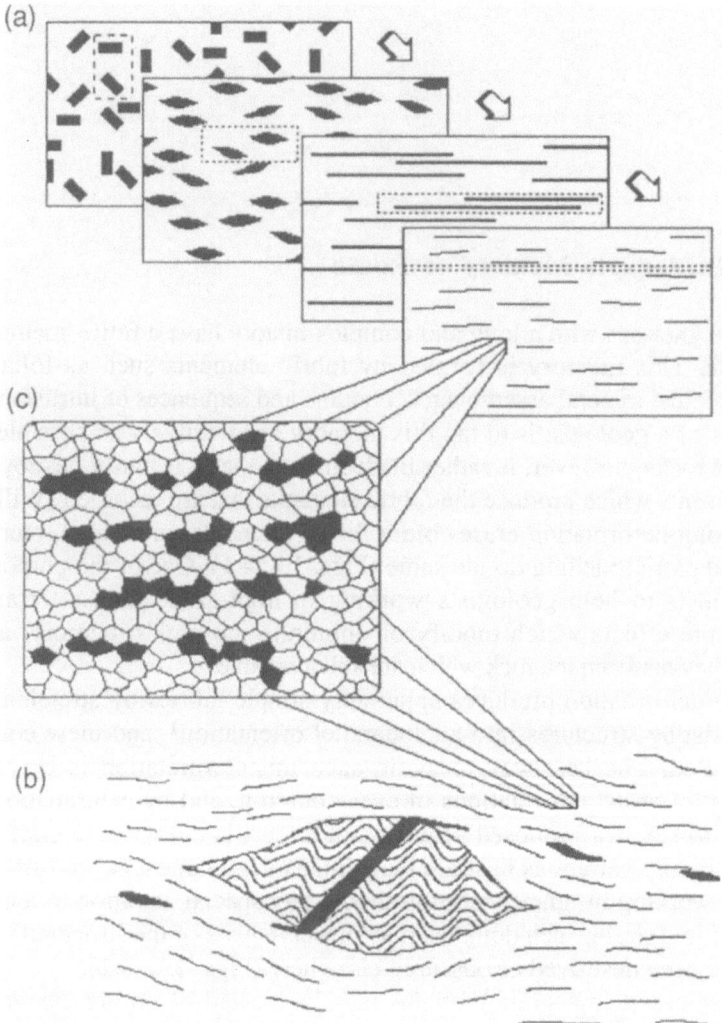

Fig. 4.1. Examples of 'memory destruction' in rocks. (a) A porphyritic granite develops into a layered rock by intense deformation, illustrated in four steps; (c) shows that recrystallisation may destroy the layering and obscure the original fabric even more, resulting in a homogeneous gneiss with a weak shape fabric. The fabric shown in (c) may also occur in rocks with a more complex history such as (b). Only close observation of the central relic lens in (b) indicates that the real history of this rock must include at least the following elements: development of a foliation - tight folding of the foliation - intrusion of a dyke - a second phase of strong deformation.

Fig. 4.2. Rootless fold of a hornblende-schist layer preserved in biotite schist, indicating strong deformation under retrograde mètamorphic conditions. Damara Belt, Namibia.

Fig. 4.3. Ductile shear zone (above) transecting a layered gneiss with flattened mafic dyke (below). The resulting structure resembles a low-angle unconformity. Habarana, Sri Lanka.

4.2 Igneous or Sedimentary Origin of Gneisses

4.2.1 General Guidelines

A major problem commonly encountered in high-grade gneiss terrains is whether a particular gneiss is a 'paragneiss' or an 'orthogneiss' (i.e. derived from a sedimentary or igneous rock, respectively). In fact, the sedimentary or igneous nature of many rocks in high-grade gneiss terrains can only be established after careful collation and analysis of *all* the structural, petrological and geochemical data that can be collected (Section 6.2).

Many high-grade gneiss terrains are dominated by rocks of plutonic origin with only a minor component derived from sedimentary rocks. This may not be immediately obvious because:

(1) Strong deformation may have rotated and distorted all the structures into parallelism. Thus the gneiss may have a monotonous planar layering and superficially resemble a little deformed metasedimentary rock (Fig. 4.3).

(2) Many igneous rocks, including granitoid rocks, were intruded as sheets parallel to an existing foliation, rather than as cross-cutting dykes or batholiths. If intrusion took place under high-grade conditions, the boundaries of these sheets may be diffuse because of partial melting of the wall rocks, and slow cooling could have led to differentiation and the formation of layering within the intrusive bodies.

(3) Deformation and igneous intrusion can produce many structures which bear a striking resemblance to primary sedimentary features.

Regularly layered gneisses are of sedimentary, igneous or composite origin. Gneisses that possess layers of various thickness and/or composition in which the compositions are *all* in the range of normal igneous rocks, and in which there is some evidence that the rocks are not strongly deformed, are probably of volcanic origin, or in some cases may represent primary layers of a major pluton. If a gneiss terrain contains thick, homogeneous units of rock that are of igneous composition, an igneous origin is likely for those units. Layering in gneisses derived from sedimentary rocks is generally more irregular in both thickness and composition, some layers may be of distinct sedimentary origin (quartzite and marble), and layers have more diffuse boundaries than in gneisses derived from igneous rocks. Thick units of high-grade metasedimentary rocks can also be homogeneous, for instance a metagreywacke, but they may be recognised by a composition either too siliceous (generally quartz-rich) or too aluminous (generally cordierite and/or biotite-rich) to be of normal igneous origin.

Care should be taken not to extrapolate the igneous or sedimentary nature of a thin vein or layer to the entire volume of rock investigated. Networks of thin quartzo-feldspathic pegmatite veins are widely developed in granitic rocks, volcanic rocks and pelitic metasedimentary rocks and do not indicate any particular origin for their host rocks. The occurrence of thin layers of quartzite, marble or other material of distinctly metasedimentary origin within a sequence of, for example, layered quartzo-feldspathic gneiss, is not proof of a metasedimentary origin of the entire sequence. Metasedimentary rocks can become tectonically interleaved, even as very thin layers, within gneisses of igneous origin, and basement/cover unconformities can be obliterated by the kind of deformation that is widespread in gneiss terrains (Fig. 3.14). It is possible that superficially similar groups of gneisses have significantly different geological histories.

4.2.2 Apparent Sedimentary Structures

In previous sections we have shown how gneissose layering that developed by deformation of a network of metasedimentary and igneous rock units may resemble bedding. Other features can be equally difficult to interpret.

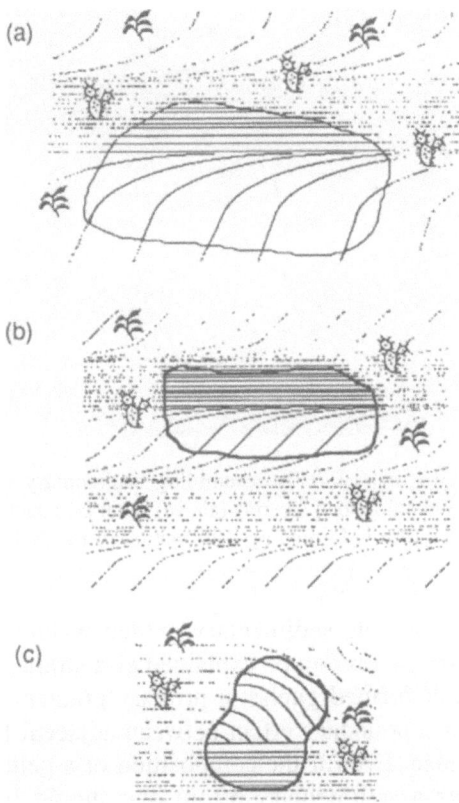

Fig. 4.4. Sketch map showing false cross-bedding where younger deposits hide crucial parts of complex deformation structures such as (a) shear zones with sharp boundaries transecting an older layering; (b) a tectonic lens of low strain preserving older layering in a major shear zone; (c) a ramp structure on a thrust shear zone. The outcropping parts of the structures are shown by solid lines.

Cross-bedding structures can survive in rocks of high metamorphic grade if they are not strongly deformed (Myers & Williams, 1985; Fig. 4). Strong deformation generally destroys cross-bedding structures but it may also generate false cross-bedding, such as:

(1) Shear zones with sharp boundaries which cut an existing layering (Figs. 4.4a; 4.5).

(2) Lenses of weakly deformed material in a shear zone that are partly hidden by other deposits (Fig. 4.4b).

(3) Small ramps in layering. Such structures may develop in gneiss along narrow shear zones or melt pockets which act as decollement planes (Fig. 4.4c).

Fig. 4.5. Deformed layering, superficially resembling sedimentary cross-bedding. Metagabbro with relic igneous mineral-graded layering (above) is cut by a ductile shear zone (below) in which a new tectonic layering is formed by deformation of the metagabbro. Fiskenæsset, SW Greenland.

Structures which resemble sedimentary graded bedding can be found in many gneisses. Such structures can be explained as a strain gradient normal to layering in a strongly deformed gneiss; a primary gradient in grain size in a sheet-like intrusion, or a boundary effect between adjacent layers of different composition or grain size. Even if the composition of a pelitic rock in a high-grade terrain indicates a sedimentary origin one should be cautious in the interpretation of relict sedimentary structures. Metamorphism of weakly deformed, primary graded-bedding structures in such pelitic rocks may lead to a reverse grading in which metamorphic minerals increase in size upwards (Fig. 4.6). This phenomenon develops under medium-grade metamorphic conditions and may persist to high-grade.

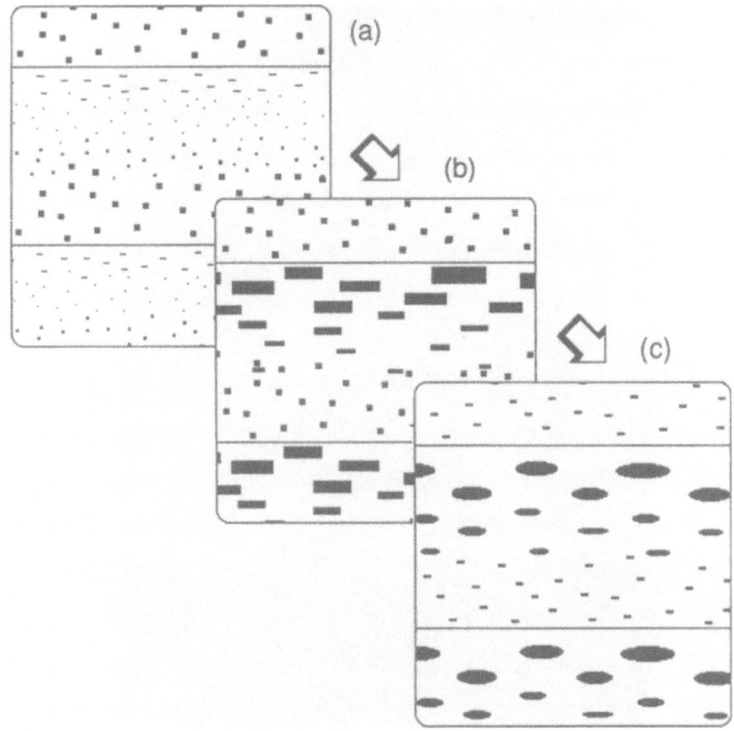

Fig 4.6. A possible sequence showing the reversal of sedimentary fining-upward graded bedding to coarsening-upward graded layering during metamorphism: the original structure (a) is overgrown by Al-rich porphyroblasts which grow to maximum size in the Al-rich pelitic top of the beds (b): subsequent deformation (c) may preserve this reversed sequence but obscure the porphyroblastic origin of the Al-silicates.

Intense boudinage or fragmentation of layering in shear zones can produce structures which superficially resemble conglomerates. These structures can usually be distinguished from true conglomerates by the presence of pseudo-pebbles of uniform composition, shape and orientation. Rocks resembling deformed quartz-pebble conglomerates can be formed by folding, boudinage and transposition of quartz veins with a variety of primary orientations into sub-parallelism during deformation. Some pseudotachylyte veins (Section 3.5.2) in gneiss terrains develop extensive layers with rock fragments which appear as rounded 'pebbles' in a dark, glassy matrix on some outcrop surfaces. Investigation of the 3D shape of the veins usually reveals their true nature.

Fig. 4.7. Isoclinal fold of a mafic dyke in orthogneiss. If such a fold closure is obscured, the layering may be erroneously interpreted as a little, or undeformed, primary stratigraphic sequence. Jequié Complex, Brazil.

4.3 Assessment of Strain Intensity

High-grade gneisses are not richly endowed with strain markers and strain values in such rocks tend to be underestimated. The following criteria can be used to recognise highly deformed rocks:

(1) Presence of boudins and isoclinal folds or rootless folds in planar, sub-parallel layering (Figs. 4.2; 4.7).

(2) Subparallelism of veins and layers which share a common foliation or lineation. Deformed veins should not be confused with a set of parallel intruded, undeformed veins or dykes (Fig. 4.8).

(3) Presence of augen or lenses of less deformed material, such as augen comprising single feldspars or weakly deformed rock fragments (Figs. 3.12; 4.1). Care should be taken to distinguish augen originating by deformation from (rare) augen formed by new growth of single crystals of feldspar.

(4) Flattened offshoots from dykes and very small angles between cross-cutting dykes.

(5) Flattened markers such as xenoliths parallel to a foliation.

4.4 Shear Zones

4.4.1 Recognition of Shear Zones

Narrow shear zones can be recognised in gneisses because they show a strain gradient from strong deformation in the centre of the zone to a weakly or undeformed wall rock (Watterson, 1968; Ramsay & Graham, 1970; Ramsay, 1980; Figs. 3.16; 4.9c; 4.11). In an undeformed host rock, the strain gradient results in a characteristic pattern of increasing intensity of foliation towards the shear zone and the bending of this foliation into the zone (Figs. 3.16; 4.9c). Pre-existing veins, layering and foliation in the host rock are displaced along such shear zones (Fig. 4.9a,b,d). Wide shear zones exceeding the size of outcrops are more difficult to recognise. Besides the high-strain criteria given above, the following fabric elements are characteristic of shear zones:

(1) An unusually regular layering of constant thickness (Figs. 3.9; 3.10e; 3.11d; 4.10).
(2) A straight linear shape fabric in the plane of the layering (Fig. 3.18).
(3) The presence of isoclinal and sheath folds in the layering with fold axes subparallel to the lineation (Figs. 4.10; 4.11).
(4) The presence of fabric elements with consistent monoclinic shape symmetry (Section 4.4.5; Figs. 3.5; 4.11).

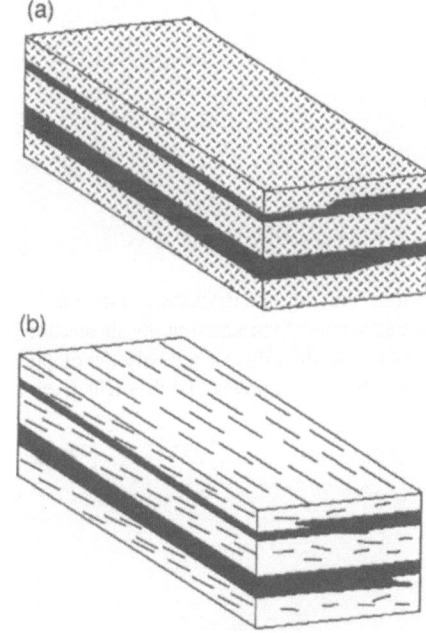

Fig. 4.8. (a) Host rock intruded by thin dolerite dykes, both undeformed. (b) Strongly deformed host rock with flattened and folded, originally thicker dolerite dykes. The situation in (b) could be mistaken for an undeformed rock such as (a) if only casual observations are made on outcrop surfaces parallel to the lineation (left face). Observations on surfaces normal to the lineation (front face) will reveal the presence of isoclinal folds in the dolerite in (b).

56

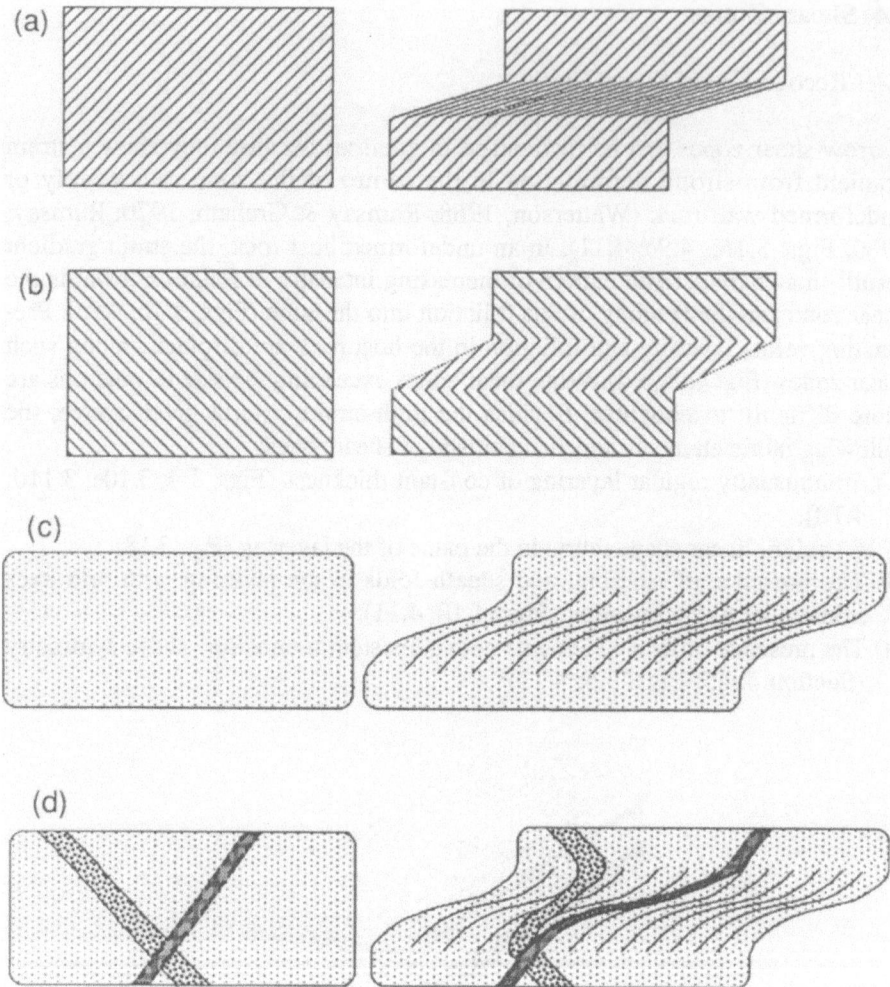

Fig. 4.9. Diagrammatic sections showing the deformation of older fabric elements by a shear zone. Pre-existing foliations will become reoriented in the shear zone, and their new orientation may be steeper (b), or shallower (a), than that of a newly developed foliation in structureless rock at the same finite strain (c); the same applies to dykes which are cut by shear zones (d).

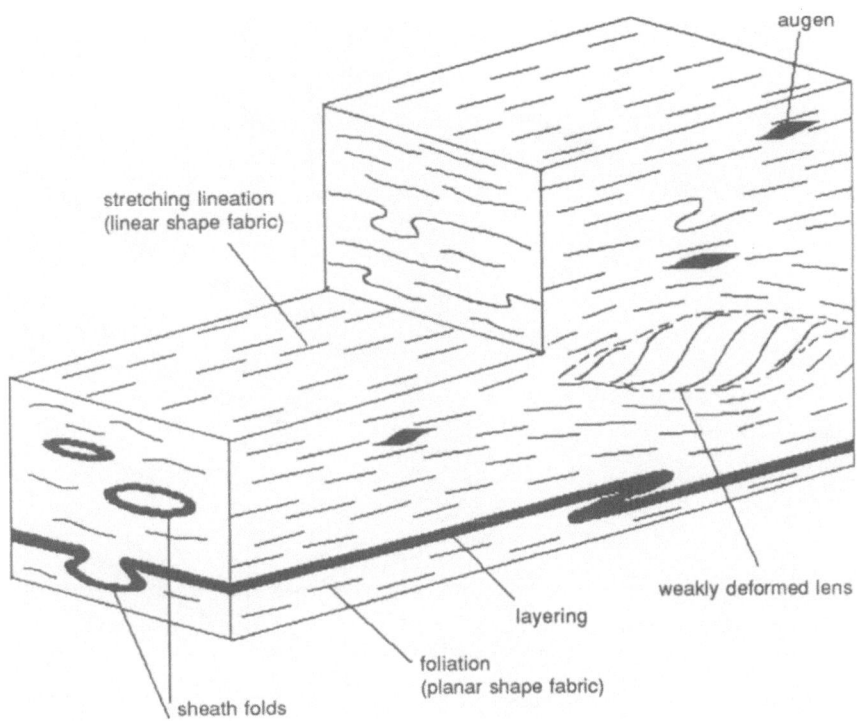

Fig. 4.10. Schematic representation of commonly developed fabric elements which can be distinguished in shear zones in the field.

58

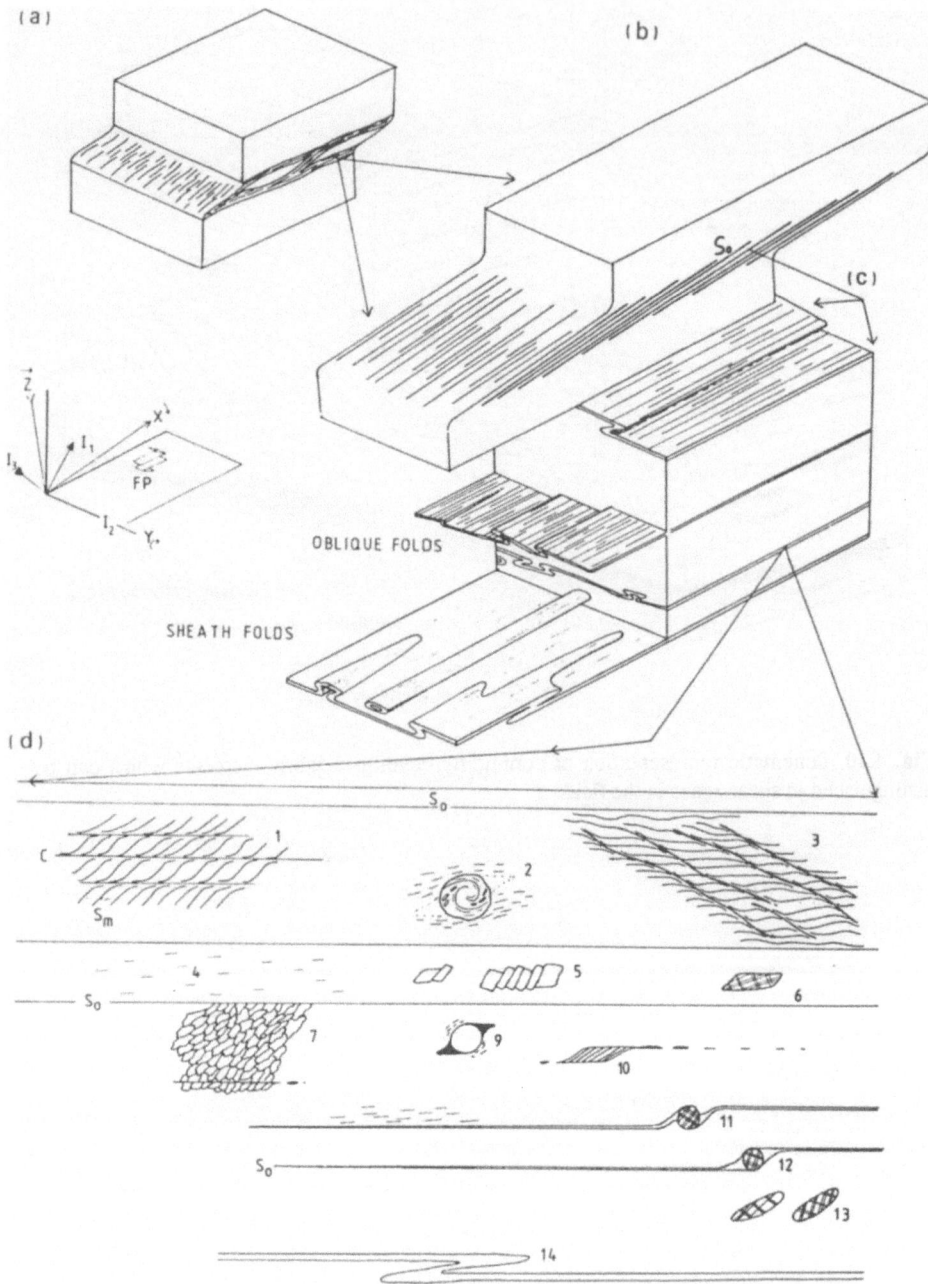

Fig. 4.11. (a) Ductile shear zone with anastomosing internal structure; (b) shear zone branch from (a) showing compositional layering (So) oblique to shear zone margin; (c) detail of (b) showing two types of isoclinal folding; (d) schematic summary of commonly seen monoclinic fabric elements (1-16) in mylonites on sections parallel to the stretching lineation (L) and normal to the compositional layering (So). Explanation in Section 4.5.5. Sm - mica preferred orientation. FP - flow plane of simple shear. After Passchier (1986a).

4.4.2 Deformation in Shear Zones

Once a shear zone has been recognised, its role in regional tectonics should be determined. It is necessary, however, to give a warning at this stage. Most shear zones are not straight uncomplicated domains of progressive *simple shear*; they may have a complex, branching structure, and terminate laterally. Flow in the branches was probably of a general non-coaxial type (Fig. 3.3, centre) rather than simple shear and the wall rock may have been deforming while the shear zone was active. Nevertheless, if the direction and sense of movement along a shear zone can be determined, its role in regional tectonics can be sufficiently assessed for most geological purposes.

4.4.3 Determining Movement Direction Using Lineations

Lineations can help to determine the relative movement direction of crustal segments which were displaced along a ductile shear zone. In brittle fault zones, slickensides of quartz fibres can be used to the same purpose. The most commonly encountered lineations in shear zones are stretching, mineral and intersection lineations. Stretching and mineral lineations are commonly parallel to the X-axis of the finite strain ellipsoid. In a shear zone where progressive deformation was by non-coaxial flow, the X-direction of the finite strain ellipsoid indicates the displacement direction of the less deformed blocks that are separated by the zone. However, the interpretation of lineations is relatively difficult and presents a large number of pitfalls for the unwary:
(1) Stretching lineations are *not* necessarily parallel to the X-direction of the finite strain ellipsoid if a shape fabric of some sort was present before the onset of deformation (an older deformation fabric, or a fluid-flow fabric in an intrusion).
(2) Mineral and stretching lineations are easily confused. Such lineations can have the same kinematic implications, but they may have different implications for metamorphic conditions during their development, so they should be identified and named correctly.
(3) Many shear zones are transected by thin brittle faults which follow the foliation; slickensides on such faults could be erroneously interpreted as stretching lineations; they can be recognised as slickensides because they only occur on the fault, not throughout the rock.
(4) If planar layering in a gneiss is subject to significant shortening during non-coaxial flow in a shear zone, this leads to foliation development which may cause the layering to split up into rod-shaped segments (Fig. 4.12a). Buckle folds can develop with fold axes parallel to the rodding. This linear shape fabric is technically an intersection lineation but it could easily be confused with a stretching lineation (Fig. 4.12b). However, its long axis is perpendicular to the X-axis of the strain ellipsoid. Such structures, if incorrectly interpreted, will cause a 90° error in the bulk movement direction

60

determined from the fabric (Fig. 4.12). Stretching lineations may be distinguished in most rocks from intersection fabrics, by their more constant orientation over a large area and by the association with sheath folds.

(5) A lineation may represent only the last movement on a shear zone; several phases of shear zone activity with different movement direction may have been present.

(6) Confusion of linear and planar structures is possible in massive gneisses.

It will be clear that lineations should be carefully observed, described, measured and interpreted before they are used in any large scale tectonic reconstructions. If possible, movement directions indicated by lineations should be checked against those identified by displaced markers.

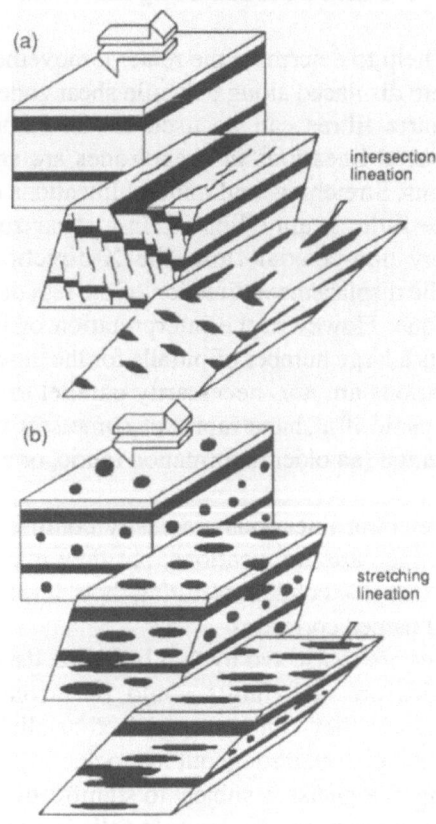

Fig. 4.12. Linear shape fabrics can develop during a non-coaxial flow by two processes: (a) the development of a foliation can form an intersection lineation normal to the shear direction by splitting existing layering; (b) mineral aggregates can be stretched into elongate ellipsoids parallel to the shear direction. It may be difficult to distinguish both types of lineations in highly strained rocks.

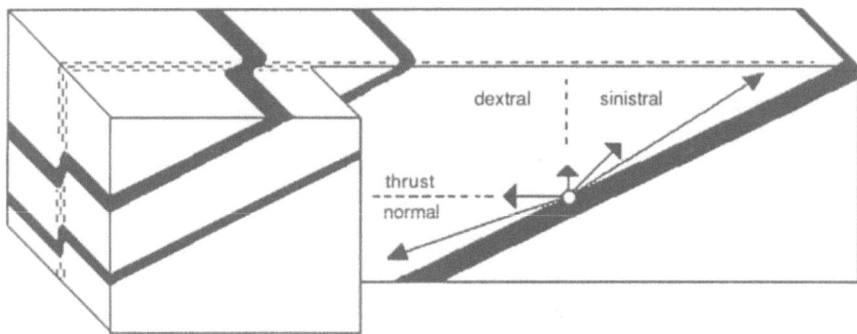

Fig. 4.13. Observations of deflected layering in a shear zone can only be used as a sense-of-shear criterion if seen on a surface parallel to the movement direction. The pattern seen on the two outcrop surfaces of the block (left) could be formed by any one of the movement directions shown on the right; both sinistral and dextral movements are possible, as are thrust and normal movement components.

4.4.4 Determining Movement Direction in Absence of Lineations

In the absence of lineations it can be very difficult or even impossible to determine the movement direction in shear zones. In fact, only the displacement of a **line** such as the intersection of two dykes which is transected by the shear zone, gives sufficient information. For example, even if a single displaced layer and the shear zone are completely exposed in 3D, and if deflection of the layer is visible, the movement direction can only be determined within an arc of 180°. In Fig. 4.13, the actual movement can be dextral extension, dextral thrusting, or sinistral thrusting. Dextral displacement of an inclined layer as observed on a horizontal outcrop surface may be the result of a dextral *or sinistral* horizontal displacement component in 3D. Only in special geometric situations, such as in Fig. 4.14, can more specific information be obtained. In this case a single horizontal displacement component cannot create the geometries seen; a steep component of movement (dip-slip) must therefore also be present.

An associated problem lies in the assessment of the amount of displacement along a shear zone if the movement direction is unknown. If the movement direction is oblique to a plane of outcrop, an insignificant deflection of markers in outcrop may represent a large amount of displacement.

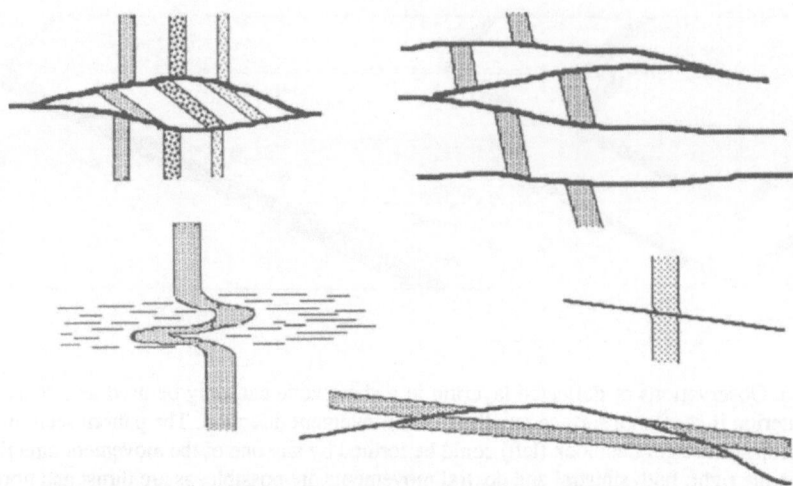

Fig. 4.14. Fabric patterns on horizontal planar outcrop surfaces which indicate that the faults or shear zones involved have a significant dip-slip component of movement.

The Use of Isoclinal Folds

If distinct lineations are absent from a mylonite with rigid wall rocks, the orientation of fold axes of isoclinal folds can occasionally be used to determine the movement direction. As explained in Section 3.4.3, both sheath folds and oblique folds in mylonites develop in such a way that their fold axes line up with the movement direction in the shear zone. Folds which were in the course of development, however, may still be at a high angle. Measurement of a large number of fold axes of isoclinal folds will usually show a clear maximum. In many gneiss terrains it may be difficult to measure fold axes if the isoclinal folds do not weather out sufficiently. In such cases, a rough impression of the movement direction can be gained by a study of outcrop surfaces with different orientations, normal to the main foliation of the shear zone (Fig. 4.10). Within shear zones, outcrop faces parallel to the movement direction may show isoclinal folds with very long limbs and a planar foliation, whereas those normal to the movement direction may show numerous tight or isoclinal folds with short limbs or tubular sections and wavy foliation planes (Fig. 4.10).

4.4.5 Shear Sense Criteria

Once the movement direction in a shear zone has been determined, the sense of movement (sense-of-shear) must be established from the fabric in the zone. Since fabric symmetry will reflect flow symmetry, observations on 'sense-of-shear markers' should only be made on outcrop surfaces normal to the foliation, and parallel to the established movement direction (Figs. 4.10; 4.11). In practice, this is usually an outcrop surface parallel to the stretching lineation in the shear zone.

In shear zones which developed under low- to medium-grade metamorphic conditions, a large number of sense-of-shear markers are available, and most are empirically established (Bouchez et al., 1983; Simpson & Schmid, 1983; Passchier, 1986a). Although most of these are only visible in thin section, some can be seen in the field, and the whole range is therefore described here (Fig. 4.11; bold numbers refer to parts of this figure) :

(a) Systematic obliqueness of shape-fabric elements or crystallographic preferred orientation patterns of constituent minerals in a mylonite with the boundaries of the zone (Fig. 4.11b: Ramsay & Graham, 1970; Simpson & Schmid, 1983).

(b) Systematic obliqueness of planar fabric elements which developed synchronously in the shear zone, such as the obliqueness of elongated newly recrystallised quartz grains (7; Means, 1981; Lister & Snoke, 1984) or a mica-preferred orientation (S_m) (4), relative to the mylonitic shape fabric S_0. This obliqueness results from the fact that some fabric elements (elongated new quartz grains, micas) rotated through a smaller angle than S_0 during non-coaxial progressive deformation.

(c) Obliqueness of S_m or S_0 with a 'shear band cleavage'[1] (White, 1979; Gapais & White, 1982). There are two types: wavy, anastomosing 'extensional crenulation cleavage' (ecc) which is oblique to the zone boundaries and to the older foliation in micaceous mylonites (3; Fig. 3.23; Platt & Vissers, 1980; White, 1979), or planar distinct 'C-planes' (1) which parallel shear-zone boundaries but lie oblique to the older foliation (S). This type is known as an S-C fabric (Berthe et al., 1979; Lister & Snoke, 1984).

(d) Lozenge-shaped single crystals of mica (10; Lister & Snoke, 1984) and feldspar (6).

(e) Fragmented rigid grains with antithetic slip between the fragments resulting from extension in a ductile matrix (5; Simpson & Schmid, 1983).

(f) Rigid porphyroclasts and their dynamically recrystallised mantles which form systems with two distinct types of internal symmetry (Passchier & Simpson, 1986): δ-type (Fig.3.17), with bent 'tails' (11) or σ-type with wedge-shaped straight 'tails' (12).

(g) A difference in position of stretched tails of recrystallised material (S_0) with respect to the central porphyroclast on which they were generated: the tails are 'higher' on one side than on the other; this effect is known as 'stair stepping' (Fig.3.17; 10, 11,12; Lister & Snoke, 1984).

---

[1] A shear-band cleavage is a foliation which transects the main foliation in a mylonite, but developed during the same deformation event that produced the mylonitic fabric; it is thought to form at a late stage of the development of a ductile shear zone, when deformation changes from homogeneous flow on the scale of the zone to more localised deformation in narrow zones, which truncate the older fabric. Because shear-band cleavages develop during shear zone activity, they are useful sense-of-shear indicators.

(h) Synkinematic porphyroblasts with a monoclinic inclusion trail such as snowball garnets (2).

(j) Obliqueness of the long axis of oblong rigid objects relative to the mylonitic shape fabric (13).

(k) Symmetry of sheath folds in sections through a 'nose' of the structure (Fig. 4.11c, 14).

In shear zones which developed under high-grade conditions, only five markers are generally available:

(a) The orientation and distribution of folded, folded-boudinaged and boudinaged veins (Fig. 4.15).

(b) The symmetry of partly recrystallised augen or lenses of relatively strong material (e.g. Figs. 3.17; 4.11d).

(c) The imbrication of elongate megacrysts in porphyritic gneiss (Blumenfeld, 1983).

(d) The deflection of primary foliation in an anisotropic rock such as a granite or gabbro, and of existing foliation or layering (Figs. 3.16; 4.9).

(e) The displacement of markers such as dykes (Fig. 4.9d) or isograd patterns which predate the shear zone (Fig. 5.6).

4.5 Folds and Boudins

Some textbooks on geology leave people with the erroneous impression that folding is associated with regional shortening, and boudinage with regional extension. In fact, any flow type comprises extensional and constrictional domains, and boudins or folds can develop synchronously in layers and veins of different orientation (Figs. 4.15; 4.18). Boudinage can be of two types; it either occurs in layers of specific composition due to a competency contrast (Fig. 4.16), or in strongly foliated rocks due to an anisotropic distribution of tensional strength. In high-grade gneiss, boudin necks are generally filled with pockets of locally derived or migrated partial melt. Folds may form by buckling of competent layers, and by rotation of layer segments in response to non-coaxial flow, as in the case of sheath folds; layers need not even be in the shortening domain of flow for sheath folds to form. Folds of layering in gneiss are usually 'similar', of class 1C, 2 or 3 of Ramsay (1967). Kink or chevron folds are less abundant except in strongly anisotropic rocks such as mica schist.

Layers or veins with a similar composition but different orientation that develop folds or boudins may rotate into parallelism at high strain (Figs. 4.15; 4.18). This results in subparallel layers which may contain either folds or boudins. This situation is often seen in shear zones. A similar situation can develop where originally parallel layers or veins of different composition are deformed by permanent extension in non-coaxial flow (Fig. 4.17). Layers that have a high competency contrast with the matrix may only develop boudins; those with no competency contrast stretch without developing macroscopic structures, whereas stretching layers that have a moderate competency contrast with the matrix layers may form sheath folds (Fig. 4.17). Such sheath folds may boudinage to form rootless folds (Figs. 4.19; 4.20). Fold trains and isolated or rootless folds are widely developed in gneisses (Section 4.6.3).

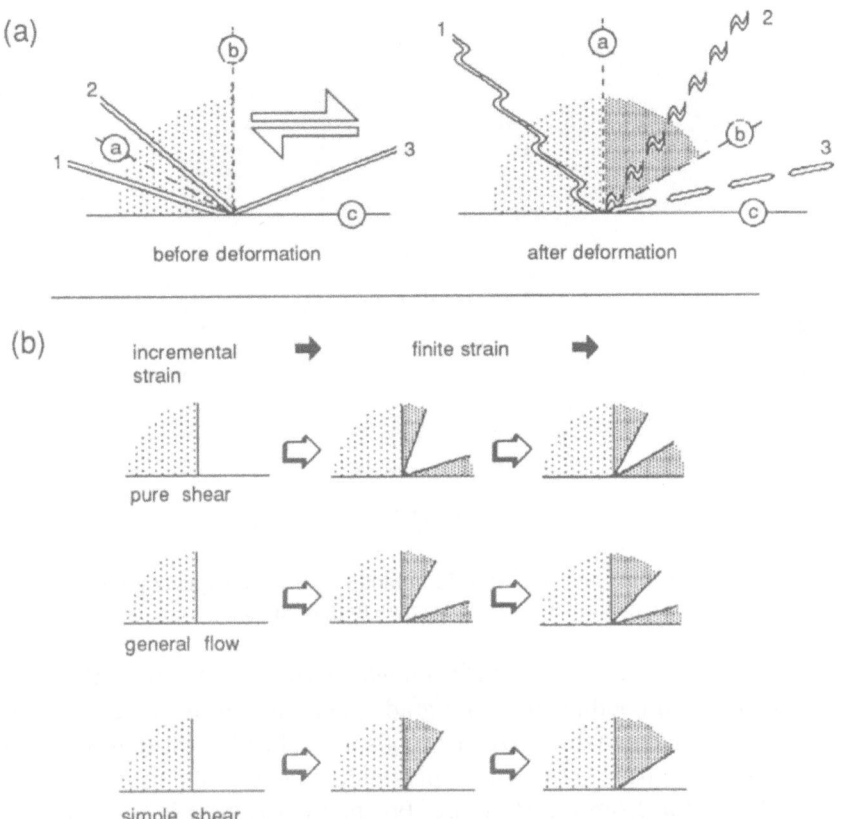

Fig. 4.15. (a) Deformation of veins by progressive simple shear. Fields are indicated of extension (white), shortening (open stippled) and shortening followed by extension (dense stippled). Depending on their original orientation, veins can become shortened (vein 1), extended (vein 3) or first shortened then extended (vein 2). Lines (a), (b) and (c) are boundaries of domains with different extension and shortening histories; (b) patterns of deformed veins for three types of progressive deformation. Note that there are no domains in which extension is followed by shortening. The diagrams are only valid for constant volume plane strain with veins normal to the plane of the drawing, and constant flow parameters.

Fig. 4.16. Boudinaged amphibolite layer (black) in biotite schist. Note quartz growth (white) and deflection of foliation into the boudin neck. Damara Belt, Namibia.

It is clear that the only direct conclusion which can be drawn from the mere presence of folds and boudins in a high-grade gneiss terrain is that a phase of deformation postdated the fabrics affected. This is a valuable contribution to the establishment of a sequence of events in the field, but one would usually like to get more information from the structure. For this purpose, it is necessary to make a detailed analysis of the orientation of fold and boudin axes and axial planes and to take note of the rock types that are involved. The relative age of veins is also important, since some may pre- or postdate part of a deformation sequence. In many cases, complex overprinting relations may exist and the following section gives an outline of possible problems and ways to handle them.

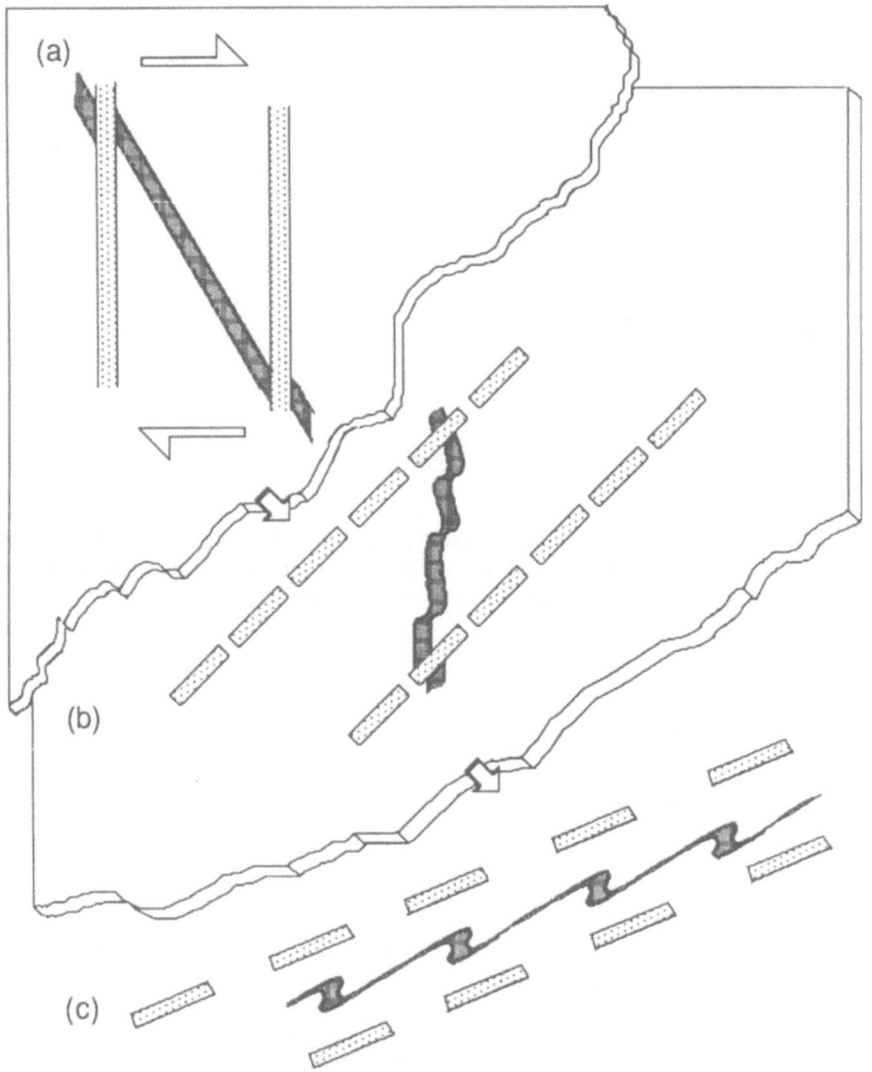

Fig. 4.17. At high strain, extension by non-coaxial flow of originally parallel layers or veins with different rheologic properties (a) may lead to parallel layers with rootless folds, boudins, and even thinned but intact and apparently undeformed layers (b and c). A fabric as shown in (c) can therefore develop by one phase of (strong) deformation and is not necessarily the result of polyphase deformation or intrusion.

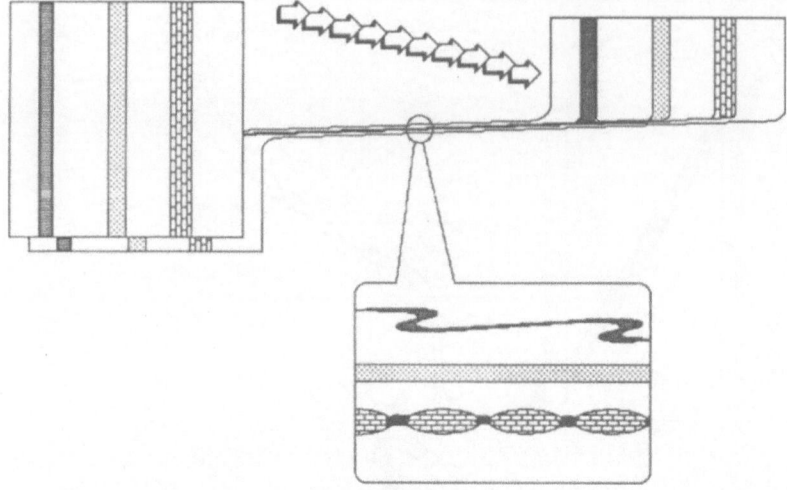

Fig. 4.18. Depending on their original orientation, veins may become folded and/or boudinaged and may rotate into parallelism at high strain; coexisting boudinaged and folded veins (bottom) may therefore belong to the same phase of deformation.

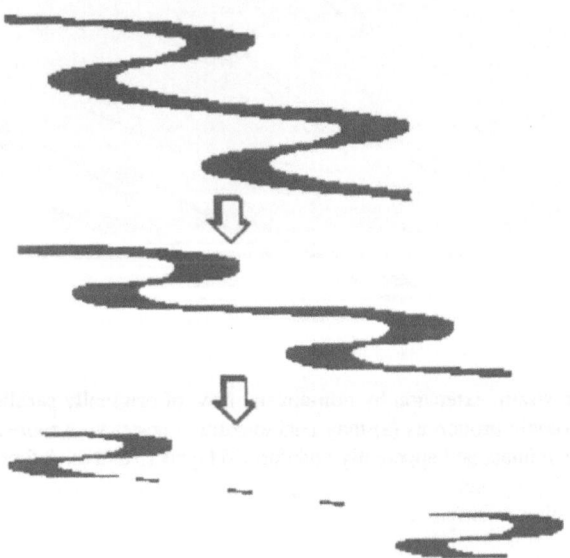

Fig. 4.19. Development of rootless folds; flattening of an isoclinally folded layer (black) which is more competent than its host rock will preferentially cause boudinage in the thin and relatively weak limbs.

Fig. 4.20. Boudinaged isoclinal (rootless) fold of anorthosite (white) and metagabbro-amphibolite (black) in intensely deformed granite gneiss (grey), with post-tectonic granoblastic recrystallisation fabric. Cape Leeuwin, SW Australia.

4.6 Overprinting Relations Involving Folds and Boudins

4.6.1 Introduction

Overprinting relations are an important tool with which to reconstruct the sequence of events in high-grade gneiss terrains. The recognition of over-printing relations is a very simple, straightforward way of dating structures relative to each other by recognition that 'A folds B' or 'A cuts B'. Despite its simplicity, many problems are associated with the overprinting argument.

One of the purposes of mapping in rocks is to derive a sequence of deformation phases which can be applied to a larger area. Such deformation phases are thought to be separated by periods without deformation. Fabric elements such as folds, boudins or foliations which formed during a certain phase of deformation may have a characteristic 'structural style'[1] and orientation in response to local metamorphic conditions, strain rate and bulk shortening direction (Hobbs et al., 1976; their Chapter 8; Williams, 1985). A regional tectonic episode may be composed of several deformation phases.

If successive phases of deformation occurred under different metamorphic conditions, the structures associated with them can be distinguished by their different structural style. If bulk shortening directions were different,

---

[1] Style is used here in the same sense as style of architecture. Style includes the geometry and mineral composition of a structure.

70

interference patterns of folds and/or boudins commonly developed which can be used to date deformation phases with respect to each other. If successive phases of deformation occurred under similar metamorphic conditions, or if they had identical shortening directions, they may have produced fabric elements of similar style and orien-tation. Such phases of deformation are difficult to separate unless some other event, such as the intrusion of veins or dykes, left an imprint on the rocks between two deformation phases.

Finally, there is also a potential problem of relative timing. Even if the rela-tive age of two overprinting structures can be determined at one locality, this age relation need not apply to a larger area. An example of this problem on a small scale is the development of sheath folds shown in Fig. 4.21. Some layers are refolded, while in other layers first folds are just developing, all during a single phase of progressive deformation.

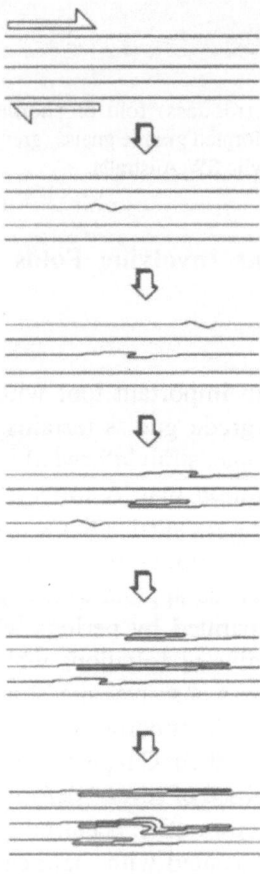

Fig. 4.21. Schematic representation of the development of sheath folds in a shear zone in which foliation is parallel to the flow plane. Sheath folds can nucleate at any time during this continuous deformation event, and so they are likely to form at different stages in the develop-ment of the final fabric. Refolding of older sheath folds is also possible during a single defor-mation event such as that illustrated here.

4.6.2 Fold Interference

Fold interference structures are usually formed by the superposition of two phases of folding (Ramsay, 1967; Thiessen & Means, 1980; Figs. 4.22 - 4.25). The interpretation of fold interference patterns is not always straightforward, and in some cases it can be difficult to detect the relative age of each phase. Such problems can be resolved if other fabric elements relating to the formation of one or both phases of folding also occur. Figure 4.26 shows an example: dome-and-basin structures that form by superposition cannot be used to determine the relative age of the folding phases (Figs. 4.23; 4.26a), unless another related fabric element such as a foliation is folded (Fig. 4.26b).

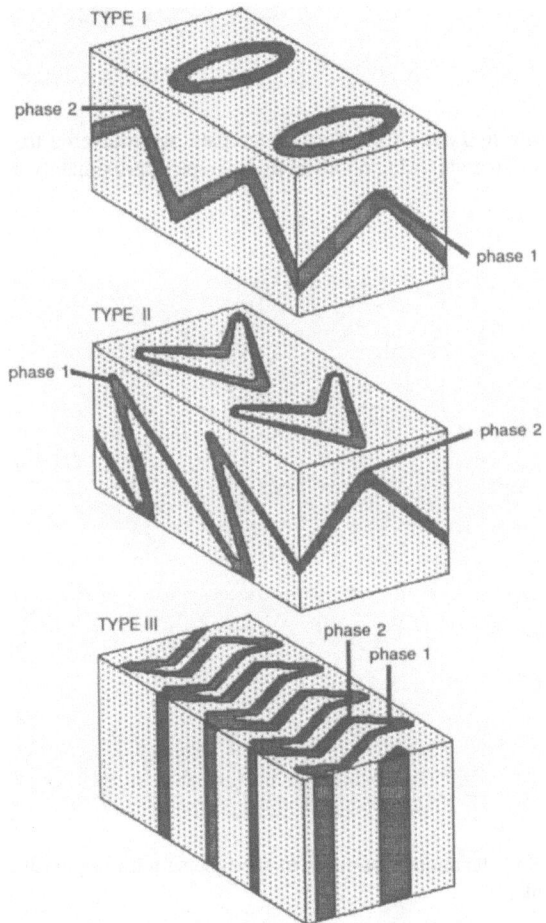

Fig. 4.22. Fold interference patterns formed by two phases of folding, defined by Ramsay (1967). Type I - dome-and-basin structures; Type II - mushroom-shape folds; Type III - refolded folds with subparallel fold axes.

Fig. 4.23. Dome-and-basin Type I-fold interference patterns in layered trondhjemitic gneiss. A late shear zone transects the fabric. Width of photograph approximately 5 m. Limpopo belt, South Africa.

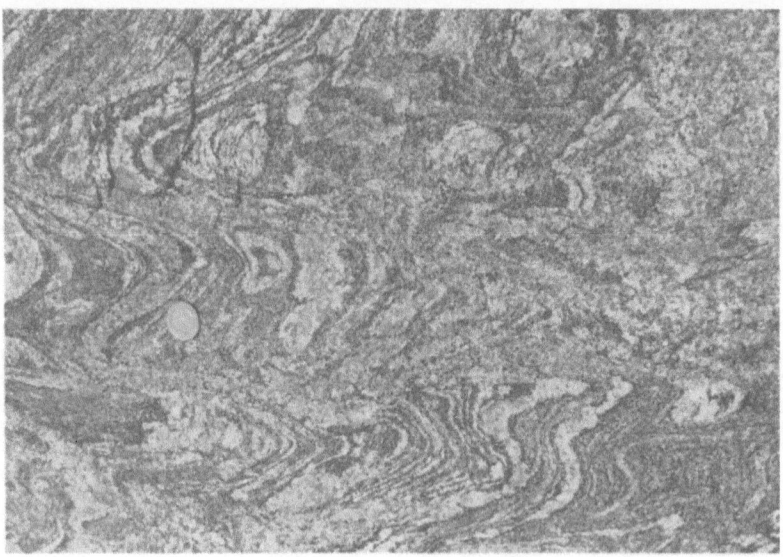

Fig. 4.24. Type II-fold interference patterns in paragneiss with migmatitic veins. Habarana quarry, central Sri Lanka.

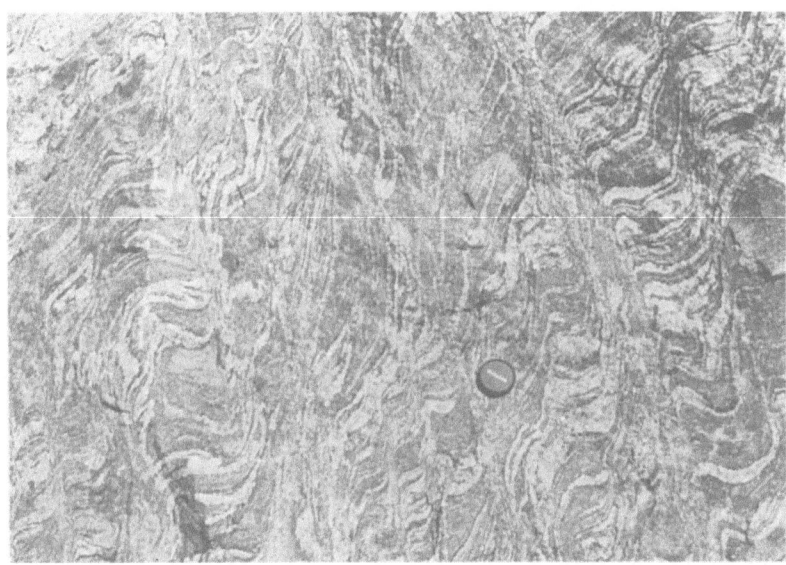

Fig. 4.25. Type III-fold interference patterns in paragneiss. The first phase of deformation produced isoclinal folds oriented east-west (right-left), and the second phase formed asymmetric folds with migmatitic veins along axial planes and locally ductile shear zones (oriented north-south). Dambulla quarry, central Sri Lanka.

It is sometimes difficult to distinguish Type I interference patterns (Fig. 4.22), formed during two episodes of deformation, from sheath folds that formed during a single deformation episode in a shear zone. Sheath folds tend to have an irregular but cylindrical shape with large segments subparallel to a stretching lineation, and they lack consistent overprinting relationships (Figs. 4.21; 4.27). Although refolded lineations and foliations can be seen in some sheath folds, they seem to reflect local events which do not represent regional phases of deformation (Fig. 4.21).

4.6.3 Deformed Vein Sets

Rootless or boudinaged folds are indicative of strongly deformed tectonic settings. They can be formed by flattening of folds which belong to an older deformation event, or during one progressive phase of deformation (Figs. 4.15a; 4.19; 4.28), as explained below.

74

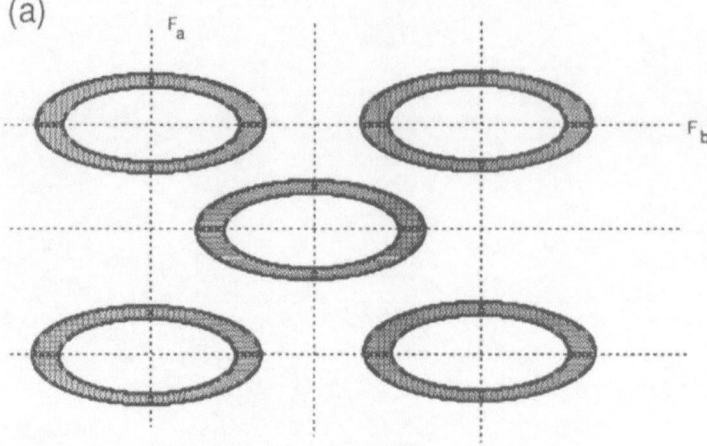

(a)

F_a

F_b

age relation Fa and Fb uncertain

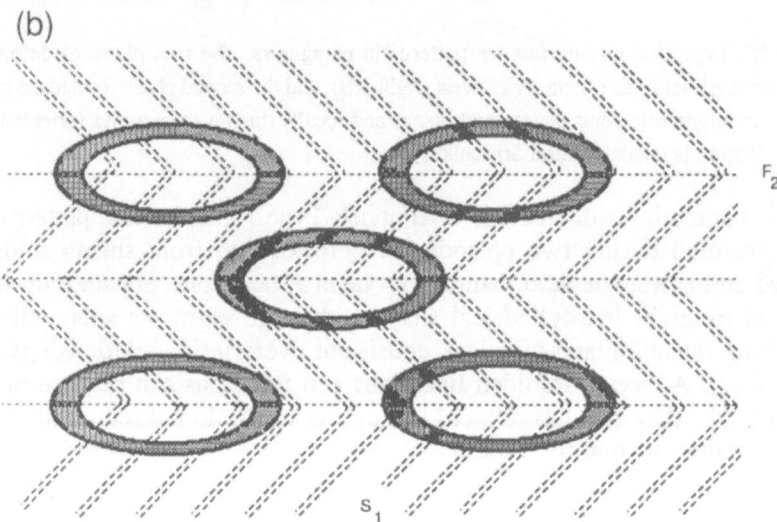

(b)

F_2

S_1

foliation (S1) older than folding (F2)

Figure 4.26. Dome-and-basin interference patterns in which: (a) it is impossible to establish the relative age of the two folding phases (Fa and Fb); (b) the relative age of the phases can be found because a foliation was developed during the first phase of folding, and was folded during the second phase of folding.

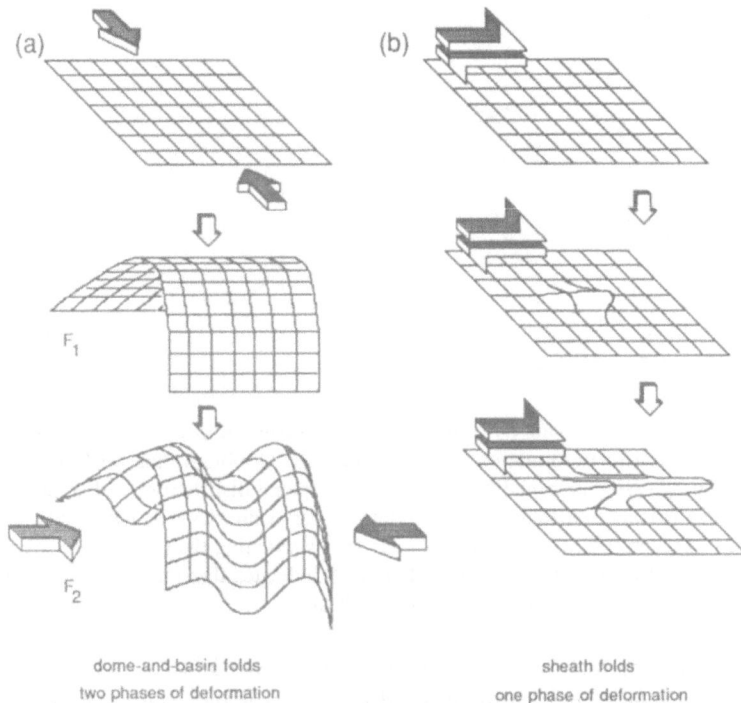

Fig. 4.27. (a) Development of a dome-and-basin fold interference structure by two phases of folding with orthogonal fold axes and axial planes; (b) development of a sheath fold by one phase of progressive non-coaxial deformation. The geometry of both structures is roughly similar.

Fig. 4.28. Folded and boudinaged granite veins in tonalitic gneiss. Barberton granite-greenstone terrain, South Africa.

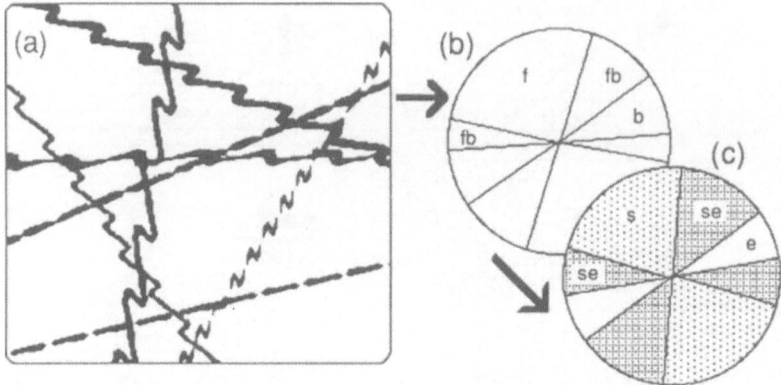

Fig. 4.29. (a) Outcrop showing deformed veins with different orientations and deformation history. The pattern of deformation types can be used to obtain data on finite strain and coaxiality of progressive deformation; (b) shows the spatial distribution of domains of folded (f), boudinaged (b) and first folded, then boudinaged (fb) veins. This pattern can be converted into patterns of extension (e), shortening (s) and first shortening then extension (se) as shown in (c). Comparison of (c) with patterns in Fig. 4.15 indicates that this example formed by a general non-coaxial flow type and dextral sense of shear.

In many gneiss terrains there are complex networks of veins with different orientations. When such vein networks are deformed, veins with different orientations will be deformed in different ways. A set of veins with various orientations in a homogeneously deforming rock will undergo extension, shortening, or both depending on their original orientations (Fig. 4.15). At any moment during the deformation, veins will lie in the instantaneous shortening or extension quadrants (Figs. 4.15a; 3.3), but they rotate with progressive deformation. This leads to some veins staying within the quadrants in which they started (1 and 3 in Fig. 4.15a), and some veins rotating from one quadrant to another (2 in Fig. 4.15a). The symmetry of the distribution pattern of shortened, extended, and shortened-then-extended veins in space (defined by lines a,b and c in Fig. 4.15a) reflects the nature of progressive deformation (coaxial or non-coaxial) and the amount of finite strain (Fig. 4.15b). It is possible to use such deformed vein sets to calculate finite strain and even to determine the coaxial or non-coaxial nature and local sense of shear of progressive deformation. The pattern of deformed veins shown in Fig. 4.29 on a planar outcrop surface, oriented normal to the veins, can be used to determine fields of folded, folded then boudinaged, and boudinaged veins. If competency contrast between veins and matrix was high, this can be translated into patterns of shortened (s), first shortened then extended (se) and extended (e) veins. The relative size of the two (se) fields is a function of the coaxiality of progressive deformation (Talbot, 1970; Passchier, in press; Fig. 4.15b).

Shortened boudins

The patterns shown in Fig. 4.15 result from plane strain, homogeneous deformation in the absence of volume change (Talbot, 1970; Passchier, in press). In other cases, the patterns become more complex (Passchier, in press). If there was a change in area in the plane of the outcrop, either by non-plane strain or by volume change, veins may also be first extended and then shortened. Such a deformation sequence may produce shortened or folded boudins (Figs. 4.30; 4.31). However, if there was no change in area, shortened boudins can only be formed by polyphase deformation. It is therefore important to note the occurrence of any shortened boudins, and to determine whether their formation was associated with a change in area in the plane of the outcrop. Shortened boudins may be the only local indication for a polyphase deformation history.

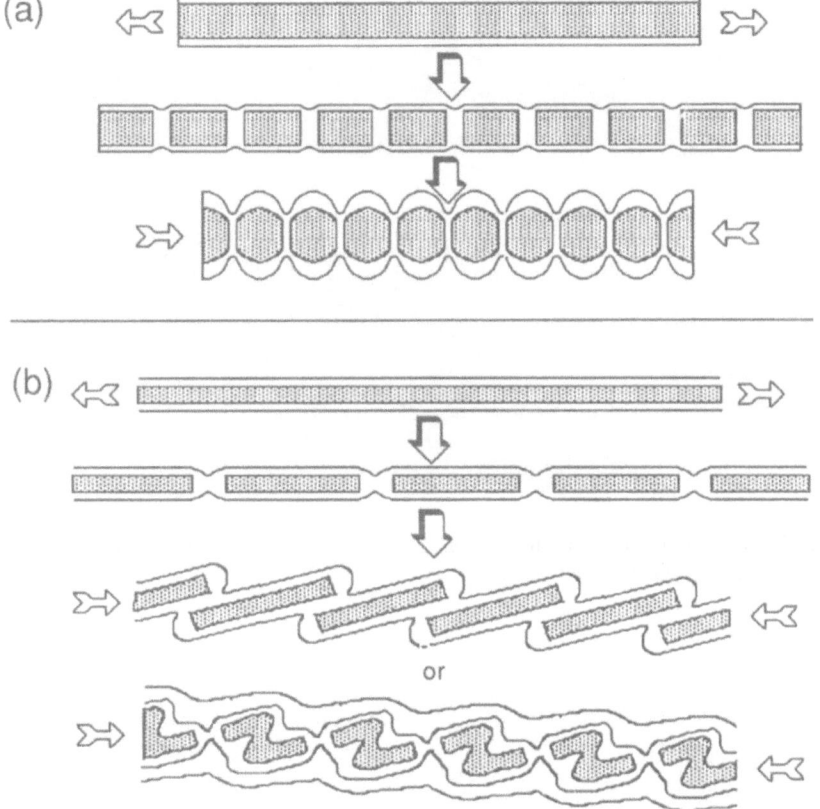

Fig. 4.30. Two types of shortened boudins; (a) semi-spherical shortened boudins are formed if original boudins were relatively equidimensional ; (b) 'rooftile' structures or folded boudins may form if original boudins were elongate.

4.7 Overprinting Relations Involving Intrusions

4.7.1 General Guidelines

Intrusive igneous rocks can play a major role in unravelling the structural and metamorphic history of gneiss terrains (Escher et al., 1975; 1976; Halls & Fahrig, 1987). Determining the relative age of foliations, lineations and cross-cutting veins and dykes may seem to be of trivial simplicity. However, it is not always straightforward (Park & Creswell, 1973) and some of the difficulties that may arise are outlined below:

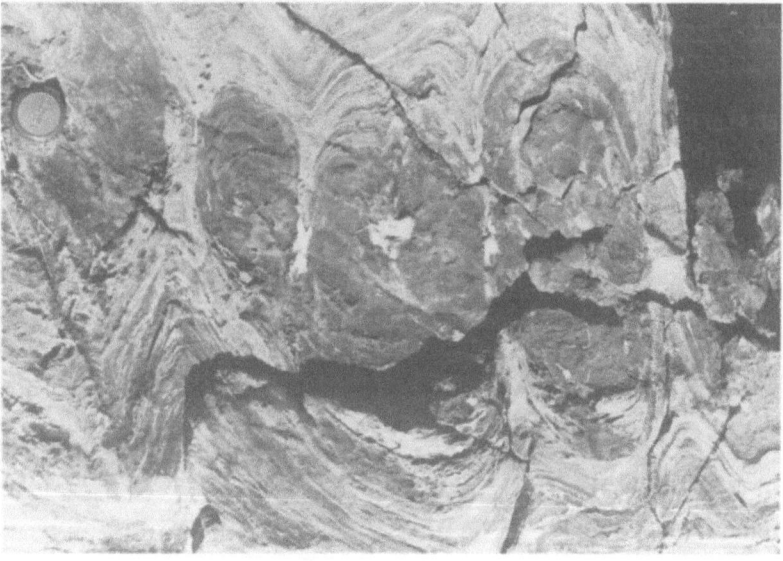

Fig. 4.31. Shortened boudins. Note the unusual geometry of folds near former boudin necks. Duchess, Mt. Isa Inlier, Queensland, Australia.

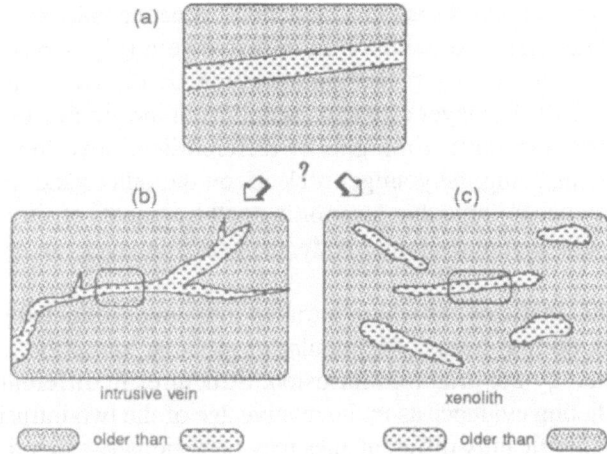

Fig. 4.32. An outcrop showing a thin layer of igneous rock in a large homogeneous granite pluton (a). The thin layer could represent a small part of either (b) an intrusive vein, or (c) a planar xenolith.

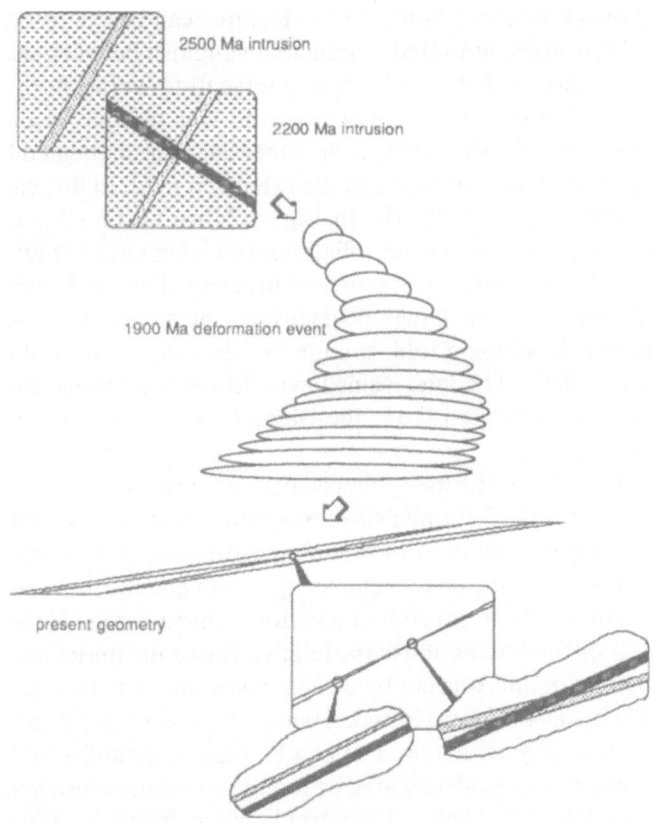

Fig. 4.33. Strong deformation can rotate originally cross-cutting, oblique intrusive veins into near parallelism. The age relationship can only be established in the field if the intersection can be located.

(1) A thin layer of igneous rock in a large homogeneous body of granite could be either a large planar xenolith or an intrusive vein (Fig. 4.32); this cannot be directly determined from the juxtaposition of the two rock types alone. However, if the thin layer crosscuts xenoliths in the granite, or has an internal zonation such as chilled margins or differentiated layering, then it can be recognised as being the younger rock. If, on the other hand, offshoots of the major body occur in the thin layer or if small fragments of similar rocks as the thin layer are seen in the larger body, then the thin layer can be identified as a xenolith.

(2) Synplutonic mafic dykes which intruded into unconsolidated granitoid rocks may break up into trains of lenticular or globular xenoliths, and/or may be back-veined by the granitoid intrusion. Structures in different outcrops will give conflicting evidence as to the relative age of the two intrusive rocks.

(3) Dykes of significantly different ages may be regionally subparallel because of strong flattening, but may locally preserve cross-cutting relations. These important cross-cutting relations may be difficult to find if outcrops are small and scattered (Fig. 4.33).

(4) In many gneisses there are straight, intrusive or partial melt veins parallel to the axial plane of folded veins. These features can be interpreted in several ways, and it is difficult to find criteria to distinguish between the alternatives (Fig. 4.34): (a) In Fig. 4.34a two intersecting veins both predate the deformation. During deformation one vein was oblique to the shortening direction and was folded, whereas the other one was attenuated but remained straight because it was stretched in the extension field. In this case both veins share a common foliation; (b) in Fig. 4.34b a second set of veins was emplaced along an axial planar foliation associated with a folded older vein; (c) in Fig. 4.34c elongate partial melt veins formed parallel to the axial plane of folded older veins; such melt pockets may spread normal to the shortening direction of a developing fold, even in the absence of an axial planar fabric (Hudleston, 1989). The later veins have diffuse boundaries and can have a weak foliation inherited from the older fabric. This situation could be confused with (a).

(5) An older, more deformed gneiss can be intruded by an apparently undeformed or little deformed sheet of granite, pegmatite or dolerite and the two rocks may then be deformed together. A strain gradient develops along the contact between the competent younger vein and the less competent older gneiss, leading to the formation of a narrow schistose zone along the contact. Hence, it is difficult to establish the relative age of the intrusions (Fig. 4.35).

(6) In Fig. 4.36 a granite was cut by a shear zone, and the shear zone itself was cut by thin dykes that were intruded whilst the shear zone was still active. If only the shear zone is exposed, it can be hard to establish the syntectonic nature of the dykes. Evidence may be found by looking closely at the fabrics; deformation fabrics are better developed in the main granite. For example, the tails on porphyroclasts in the granite are larger than those in the dyke, suggesting that the granite has suffered more deformation than the dyke.

(7) A layer of igneous material that lies within a fold, parallel to folded layering, has either been folded, or intruded after folding, following the older layering. The presence of a foliation in the igneous layer, or small veinlets branching out of it can help to decide on the actual sequence of events.

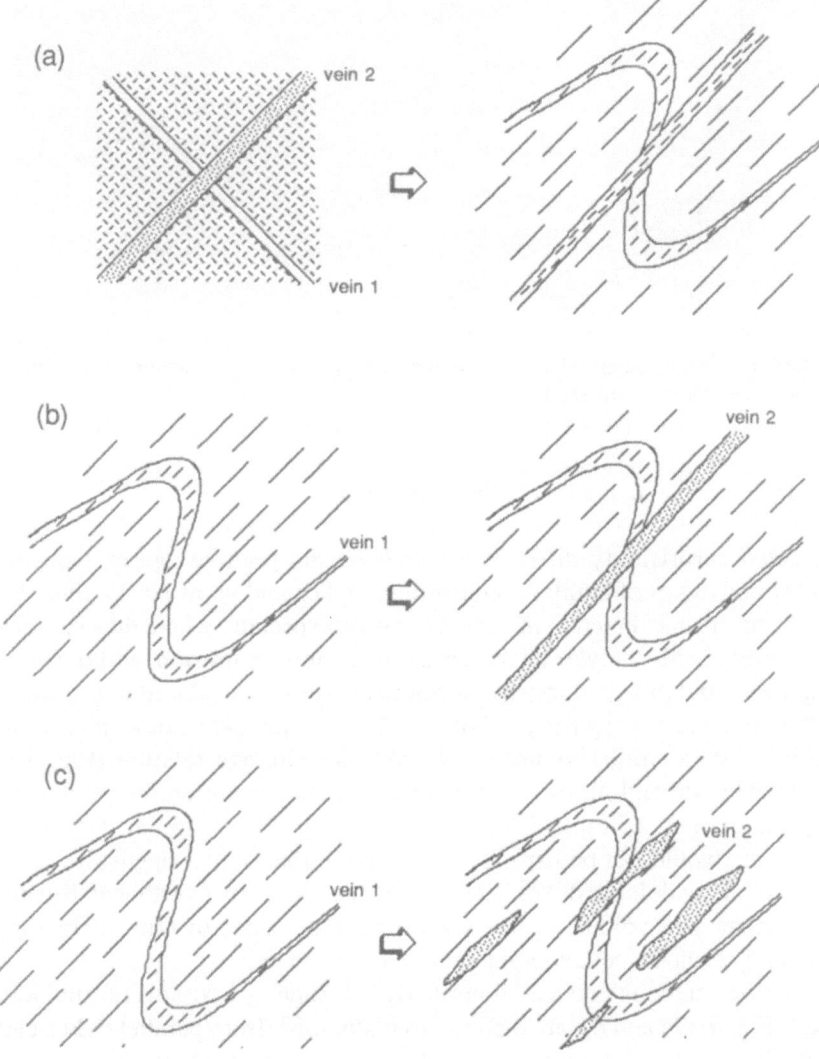

Fig. 4.34a-c. Three different sequences of events that produce a similar geometry of folded, intersecting intrusive veins.

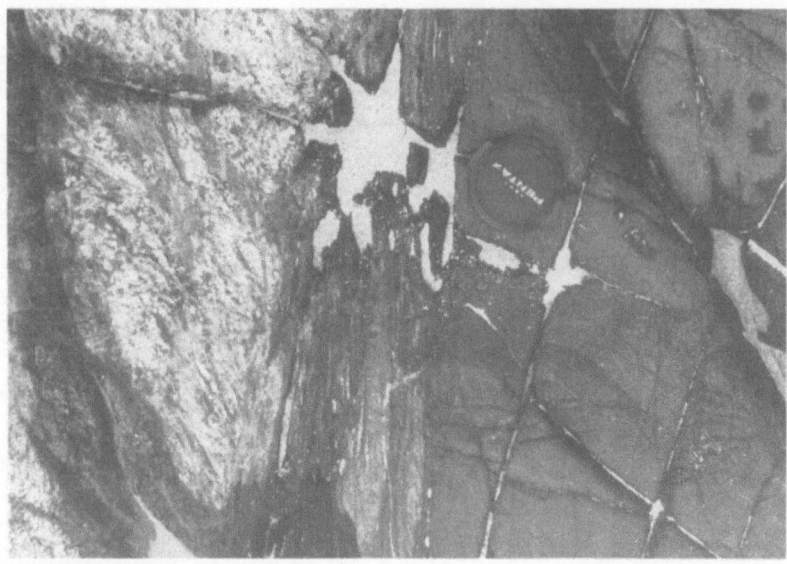

Fig. 4.35. Low-grade ductile shear zone in the contact zone of Lewisian gneiss and a dolerite dyke. Achmelvich, northern Scotland.

4.7.2 Intersection Geometry of Planar Intrusions

It is often surprisingly difficult to establish the relative age of intersecting undeformed dykes of similar composition. Dykes and veins have a tendency to branch and change direction in a complex way, especially where they cut through another dyke. Most dyke channels form in pure extension without a shear component, and this produces a characteristic 'step' in displaced non-orthogonal markers (Currie & Ferguson, 1970; Fig. 4.37a). However, some 'jogs' at intersections have a similar geometry but the opposite age relation (Fig. 4.37b). Pseudotachylyte and narrow ductile shear zones are preferentially developed along the margins of some dykes and may cause offsets similar to that in Fig. 4.37a. This would not be hard to see along the margins of pegmatite or aplite veins, but along dolerite dykes such fault rocks are easily overlooked because of their similar dark colour. In this situation the presence of fault rocks may be detected if offshoots cut into light coloured country rock.

Intersection relations between dolerite dykes, such as 'crossovers' and 'joining dykes' (Fig. 4.37c and d), are difficult to distinguish from parallel dykes cutting a fault, faulted parallel dykes or branching single dykes. However, the order of intrusion may be established by careful study of the geometry of the contacts between gneiss and dyke material, of the (generally narrow) chilled margin and, if the chilled margin is poorly developed, of truncated xenoliths (Fig. 4.37e and f). Care should be taken not to confuse chilled margins with fine-grained shear zones in this case.

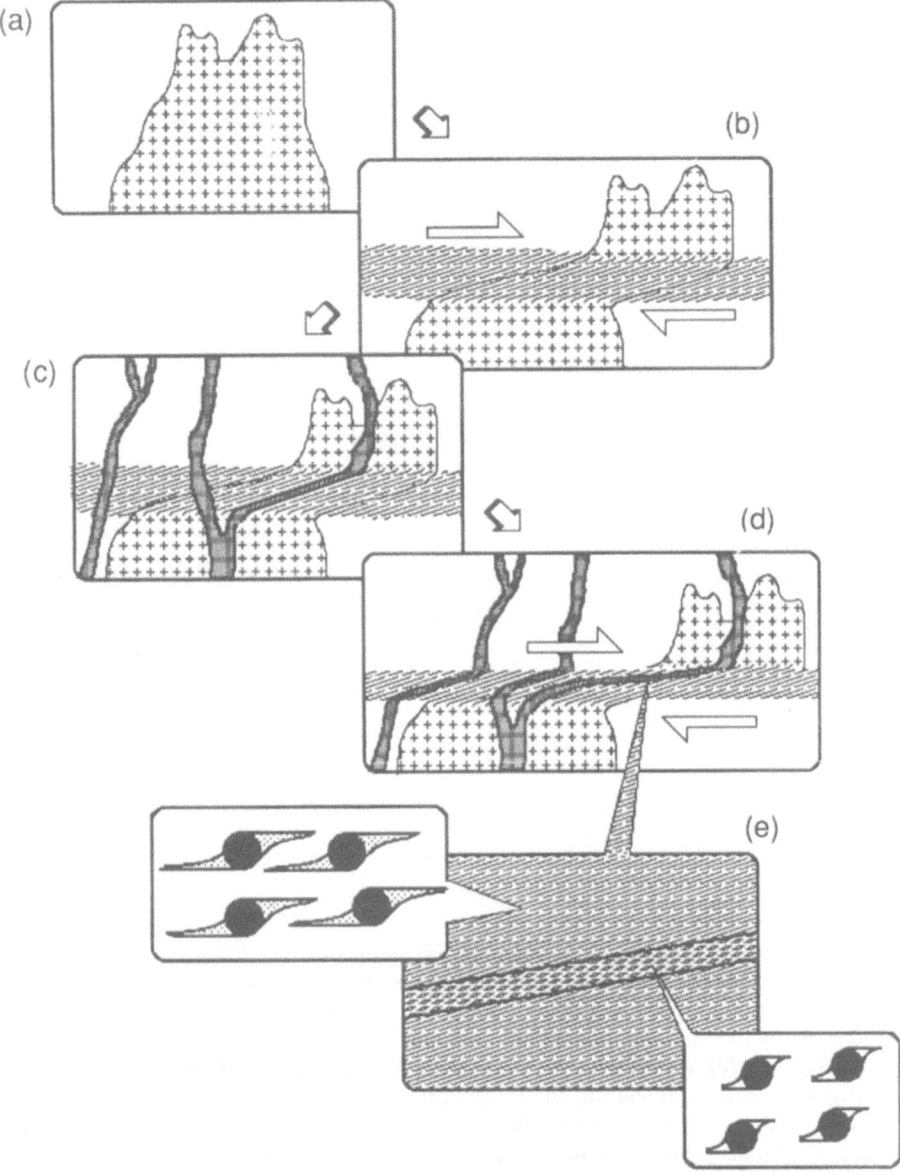

Fig. 4.36. Sketch maps showing some of the complexities of synkinematic intrusions. A granite (a) is cut by a ductile shear zone (b). During this deformation, granite dykes (black) are intruded across the shear zone (c). The dykes are deformed within the shear zone, but less so than the host rock (d). It may be difficult to recognise the different deformation states of these two rocks within the shear zone. The kind of evidence that might be seen is shown enlarged in (e); the tails on porphyroclasts in the dyke are less well developed than those seen in the more deformed host rock.

84

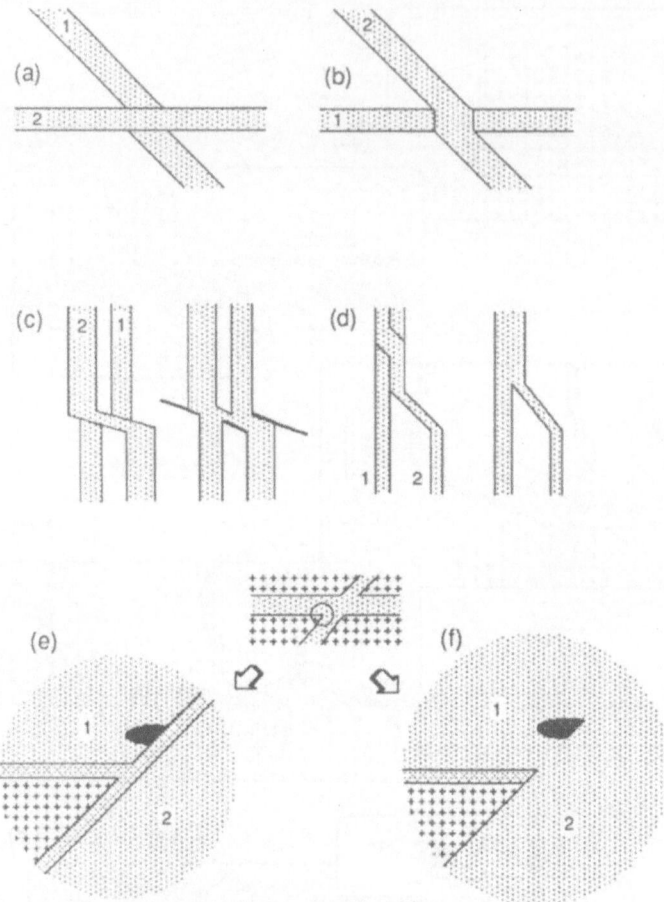

Fig. 4.37. Intersection relations between planar intrusions: (a) intrusion by dilatation without a shear component results in the distinct displacement of markers such as older dykes; (b) development of a jog where a dyke steps sideways as it cuts through an older dyke; (c) 'overstepping' by intrusion along a jog (left), and intrusion of a pair of dykes joined through an older fault (right); (d) two phases of intrusion along the same channel (left), and a branching dyke (right). Situations (a) and (b), and (c) and (d) could easily be confused without careful observation of the intersection relations. In (e) the intersection sequence of major dykes can be established from the geometry of the chilled margins (light grey), but if chilled margins are absent or obscured (f), truncated xenoliths must be used to determine the order of emplacement.

4.7.3 Interaction of Dykes and Shear Zones

The relative age of planar intrusions and shear zones is difficult to establish if both are parallel. There are two possibilities:
(1) a vein was intruded into a shear zone either during or after the deformation (Fig. 4.38a).
(2) a shear zone nucleated in or along a pre-existing vein because of the rheology contrast (Fig. 4.38b). The following arguments can be used to distinguish between these situations.

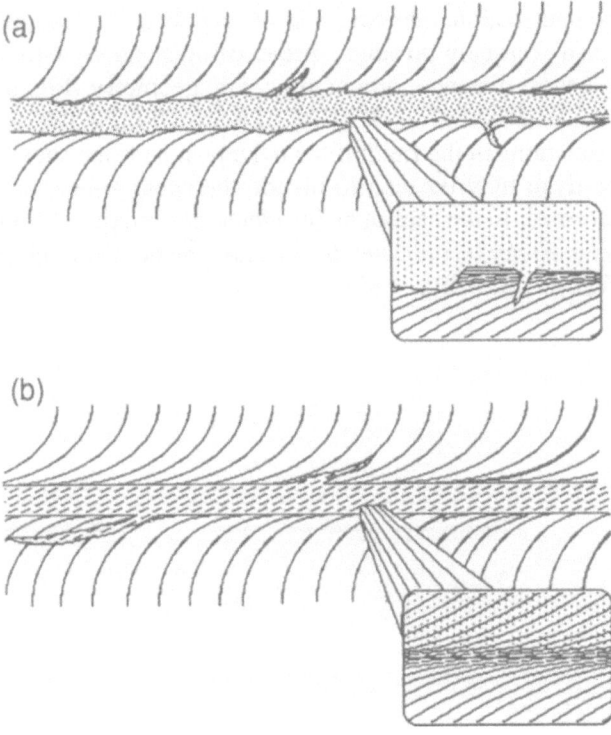

Fig. 4.38. Two superficially similar situations representing opposite sequences of events: (a) A vein was intruded along an older shear zone; it cuts the foliation of the shear zone, and displays angular jogs and branching veinlets in the margin of the zone. (b) An intrusive vein that was the nucleus of a younger shear zone may have an internal foliation and flattened jogs and branching veinlets.

A deformed vein in a shear zone must either predate or be contemporaneous with that shear zone. This can easily be recognised if the vein is folded, boudinaged or if it has developed a shape fabric. Unfortunately, deformed veins do not always show such fabric elements. If the vein were thoroughly recrystallised during and after the deformation, or if deformation in the vein was homo-geneous, its deformed nature may be obscured. In such cases the overall shape of the vein may be of assistance. If the contact is very irregular (more so than the associated shear zone) the vein must be undeformed (Fig. 4.38a); if the contacts are planar and straight, the vein may be deformed (Fig. 4.38b). Close investigation of the fabric in minor branching veinlets, and the age relation between these veinlets and the shear zone fabric may also help. Shear zones that develop along the margin of an intrusive vein will show at least some continuation of the foliation in the vein (Fig. 4.38b - inset). However, veins that are younger than the shear zone cut the foliation in the zone (Fig. 4.38a - inset).

The intersection relations of oblique planar intrusions and shear zones can also be difficult to establish. The displacement of a dyke by a shear zone or fault indicates that the shear zone or fault is younger than the dyke (Figs. 4.39a,b; 4.40), but if the movement direction of the shear zone is subparallel to the dyke/shear zone intersection, there may not be any visible displacement. A dyke that was emplaced obliquely through a pre-existing shear zone may follow the shear zone for a short distance, producing a 'jog' geometry (Fig. 4.39c) which resembles a displaced dyke (Fig. 4.39a,b). It is usually possible to distinguish between these situations in the field; if the dyke postdates the shear zone, minor branches of the dyke may be seen to invade the shear zone, and the dyke is internally undeformed. In the case of an intruding planar dyke, the relative age of dyke and shear zone may be indicated by a characteristic 'stepping' displacement of the latter (Figs. 4.37a; 4.39d; 4.41).

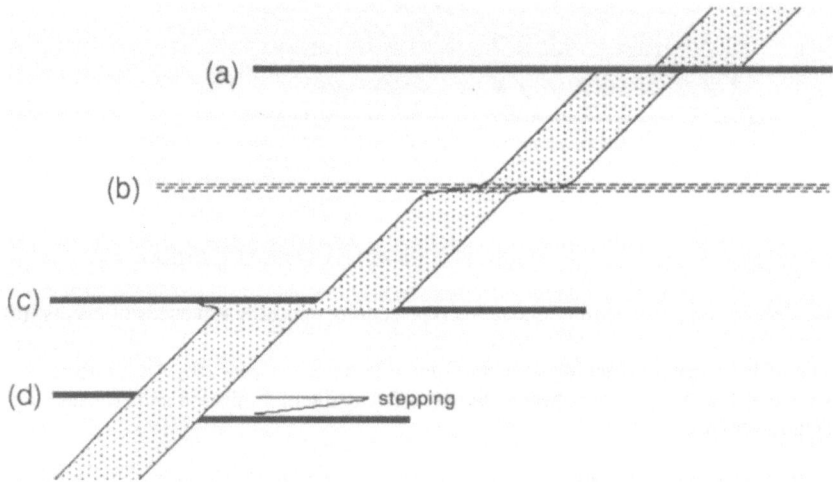

Fig. 4.39. Intersection relation between a planar intrusion and faults or narrow ductile shear zones: (a) a fault cuts an older dyke; (b) a ductile shear zone cuts an older dyke, and deflects the dyke margin; (c) a dyke cuts an older fault and forms a jog; (d) a dyke cuts an older fault without jogs, and causes 'stepping' of the fault. Situations a, b and c could be confused if the intersections are not carefully examined.

Fig. 4.40. Brittle fault cutting a dolerite dyke. Width of dyke 5 cm. Vestfold Hills, Antarctica.

Fig. 4.41. Brittle fault cut by a dolerite dyke which followed the fault for a short distance. Vestfold Hills, Antarctica.

In wide shear zones, the orientation of dykes can help to establish their age with respect to the zone, but again, care is needed. Figure 4.42 shows a straight dyke (c) cutting across a sinuous foliation in a major shear zone. In many cases,

88

however, transecting dykes follow the foliation in a shear zone, because the foliation is a strong material anisotropy which may guide the leading crack of the intrusion. Only close inspection of the dyke/shear zone contact will reveal the true nature of their relative age. Consider the situations in Fig. 4.42 where a set of parallel dykes (a) and (b) cross a shear zone. It is possible that (b) predates the shear zone and that (a) was intruded during the formation of the shear zone. However, it is also possible that (b) postdates (a) and was intruded along the foliation of an older shear zone. (a) and (b) may both predate the shear zone if (b) had a jog in the future location of the zone (due to an older brittle fault in this location). Finally, both (a) and (b) may postdate the zone if the leading tips of the intrusions (a) were only weakly affected by the anisotropy in the shear zone. The real story can only be determined by inspection of the dyke/shear zone contacts and by the presence or absence of a deformation fabric in the dykes.

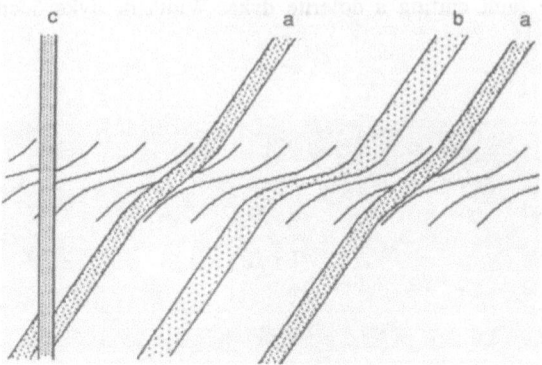

Fig. **4.42**. Intersection relationships of a planar intrusion and a wide ductile shear zone. Intrusions can transect a shear zone without deflection (c), but relations as for (a) and (b) are less obvious. The present geometry can be due to three possible sequences of events: (1) (a) and (b) intruded during development of the shear zone, but (b) intruded along a jog in the centre of the zone, causing the greater finite deflection; (2) (b) predates the shear zone and (a) intruded during shear zone activity; (3) (a) and (b) postdate the shear zone, (a) was slightly deflected by the foliation during intrusion through the zone, (b) follows the foliation in the zone.

4.7.4 Reactivated Shear Zones

Reactivation of shear zones is a common phenomenon, but difficult to recognise, except when intrusions such as dykes or veins were emplaced between multiple phases of deformation. Some splays of brittle fractures may become reactivated by ductile deformation (Fig. 4.43 a,b), and wide shear zones may be constricted to narrow zones (Fig. 4.43 c,d). If the isotopic date of such a dyke or vein could be determined, then it would provide a limit to the age of the deformation events. The mineral content and fabrics in shear zone segments that were active at different times can provide valuable information on changing metamorphic conditions.

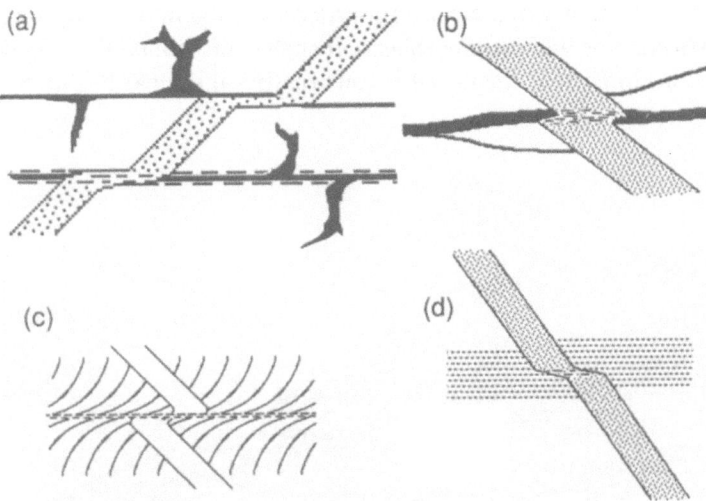

Fig. 4.43. Intersection relations of planar intrusions and polyphase shear zones. In (a) two pseudotachylyte veins (black) were cut by an intrusion (stippled), and the lower pseudotachylyte vein was reactivated as a ductile shear zone. Reactivation is indicated by the presence of a foliation in the zone and of deflected injection veins. The situation might be confused with those in Fig 4.39a and b. In (b) a branching brittle fault (black) was cut by an intrusion (stippled); one branch on each side was reactivated as a ductile shear zone, the others were not used because they were not in the same orientation suitable for reactivation. In (c) a major high-grade ductile shear zone was cut by an intrusion which was later cut by a low-grade shear zone that developed along the centre of the pre-existing zone. In (d) a wide shear zone (open stipple) was cut by an intrusion (dense stipple) that was subsequently deformed by reactivation of the centre of the shear zone.

4.7.5 Partial Melt Veins

Partial melting occurred in many gneisses that passed through prograde meta-morphism from medium to high grade and in some cases during retrogression from high grade conditions[1] (Mehnert, 1968; Ashworth, 1985). The partial melts generally formed patches and veins *in situ* or migrated to form veins nearby in boudins or tension gashes (Fig. 4.44). In-situ melt may be recognised by a patchy distribution of quartz and feldspar overprinting and erasing older fabric elements (Figs. 4.24; 4.25). Many melt patches and veins which formed under

---

[1] The word 'migmatite' is commonly used to describe a gneiss or schist with a large volume of material that formed by partial melting.

grade metamorphic conditions contain garnet that is replaced by clusters of biotite during retrogression to medium grade. Some granitoid melt patches ('leucosomes') are bounded by dark rims rich in biotite and hornblende ('melanosomes'). Although these rims are often interpreted as the residual minerals left during the generation of a quartzo-feldspathic melt, they could also have formed by retrogression of anhydrous phases when the melt veins solidified. Partial melt veins are useful criteria for distinguishing the relative timing of deformation and metamorphism. If such veins contain Fe-Mg silicates, they can be an important means of reconstructing the local P-T-time history (Chapter 5).

Fig. 4.44. Partial melt in layered gneiss in a boudin neck. If the partial melt could be isotopically dated it would also provide the age of the deformation and associated metamorphic conditions. Width of photograph 50 cm. Kandy, Sri Lanka.

4.8 Outcrop Surface and Fabric Patterns

4.8.1 General Problems

Many high-grade gneisses crop out on smooth surfaces, and in these cases structural information has to be obtained from two-dimensional planar or curved cross-sections. In many of these outcrops, the shapes and orientations of the structures can only be determined if they can be seen on three or more surfaces with different orientations. This section describes some examples.

If folds are cut by an outcrop surface at a small angle to the fold axis, relatively open folds may appear to be tight or isoclinal (Fig. 4.45a). Conversely, isoclinal folds may appear to be open on some irregular outcrop surfaces

(Fig. 4.45b). Irregular, concave and convex outcrop surfaces of planar layering may give rise to apparent folding (Fig. 4.45c) or to apparent dome-and-basin Type I-fold interference structures of Ramsay (1967) (Fig. 4.45d). Gully or ridge-like topography cutting tight or isoclinally folded layering can also produce apparent Type II or Type III-fold interference patterns (Fig. 4.45e, f).

True interference patterns change their aspect with the orientation of the outcrop surface on which they are observed and with different parallel sections through a single structure (Ramsay, 1967). Some outcrop surfaces through a doubly folded sequence may lack indications for fold interference, or even for folding (Fig. 4.22; Type III).

Very gentle folds in a layered gneiss may appear on some outcrop surfaces to define a shear zone (Fig. 4.45g). Such situations can usually be recognised because minor fabric elements do not fit a shear zone setting.

Symmetric folds may appear to be asymmetric on some outcrop surfaces, and it is necessary to combine data from outcrop surfaces with different orientations to find the orientation of the fold axes and the true shape of the folds.

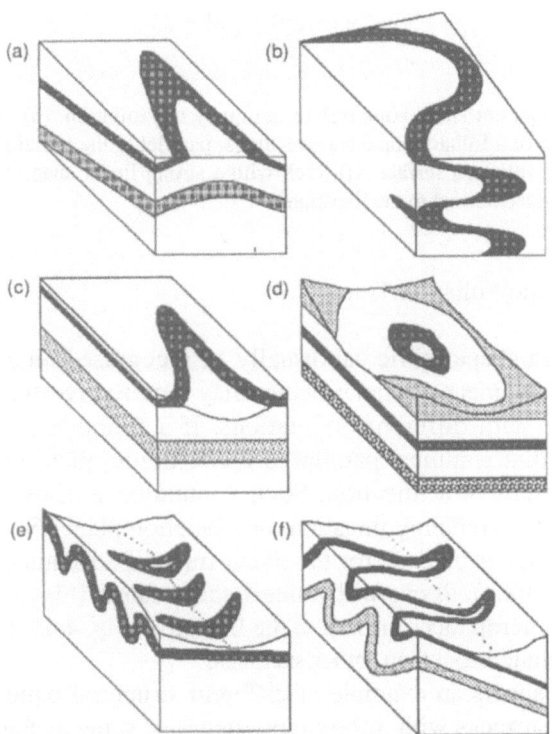

Fig. 4.45a-f. Geometric patterns as a function of the orientation of planar and curved outcrop surfaces: (a) Apparent tight folding of gently folded layering. (b) Apparent open folding of tightly folded layering. (c) Apparent tight folding of planar layering. (d) Apparent fold interference patterns of Type I in planar layering; apparent fold interference patterns of Type II (e) and Type III (f) in layering with one phase of folding.

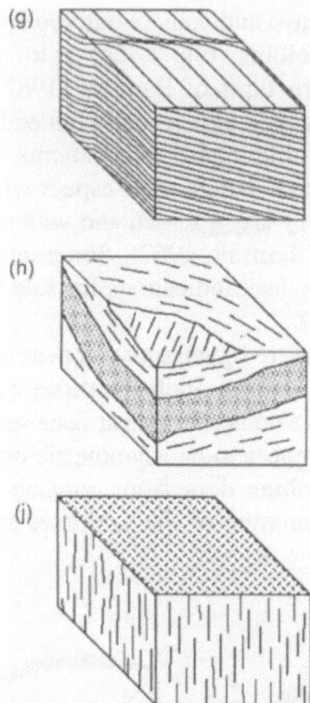

Fig. 4.45g-j. (g) Apparent shear zone due to obliquely cut foliation. (h) Apparent lineation reflecting the outline of a foliation on a surface almost parallel to the foliation, and a true stretching lineation on a foliation surface. (j) Rock with a strong linear shape fabric that appears undeformed on a surface normal to the lineation.

4.8.2 Lineations and Foliation Traces

The presence of a shape fabric can usually be recognised on any surface, but identification of its linear or planar nature may require investigation of several outcrop surfaces with different orientations. If a foliation is exposed on an outcrop surface that is almost parallel to the foliation, then the outline of the foliation may resemble a lineation. Such a situation is shown in Fig. 4.45h together with a true stretching lineation on a foliation plane. Note that the word lineation should *never* be used for the linear trace of a foliation on an outcrop surface (Section 3.2.2). Rocks with a linear shape fabric (Fig. 3.18) will appear undeformed on outcrop faces normal to the lineation (Fig. 4.45j top) and may be confused with foliated rocks on other surfaces.

Figure 4.46 shows an example of different structural patterns that can be seen on outcrop surfaces with different orientations, some distance apart, in the Revenue granite pluton near Duchess, Australia. In most outcrops (Fig. 4.46a), fold patterns are seen which could be interpreted as a folded foliation (planar shape fabric). Flattened xenoliths (black) are seen to be folded on these surfaces. However, on a few outcrop surfaces unusual 'dot-stripe' patterns can be seen.

These consist of a regular alternation of layers with a shape fabric and layers of apparently undeformed granite (Fig. 4.46b). Flattened xenoliths parallel to the shape fabric are cut by these layers of apparently undeformed granite. Such patterns could be interpreted as a foliated granite with flattened xenoliths, intruded by a swarm of granite dykes. However, if this was the case one would expect that the dykes would also be found on outcrop surfaces which show folding, either crosscutting the folds, or as folded dykes. In fact, no dykes can be seen on these surfaces, and close examination of a few outcrops with surfaces of various orientations show that the structure is simply a folded linear shape fabric (Fig. 4.46c). On outcrop surfaces normal to the fold axis, planar and linear shape fabrics cannot be distinguished; on outcrop surfaces parallel to the fold axis, 'dot-stripe' patterns are the effect of a changing orientation of the linear fabric with respect to the outcrop. The apparently undeformed layers are fold limbs in which the lineation is normal to the surface; in the other layers it is subparallel. Clearly, a wrong conclusion could be drawn in this case if data from different outcrop surfaces were not carefully compared.

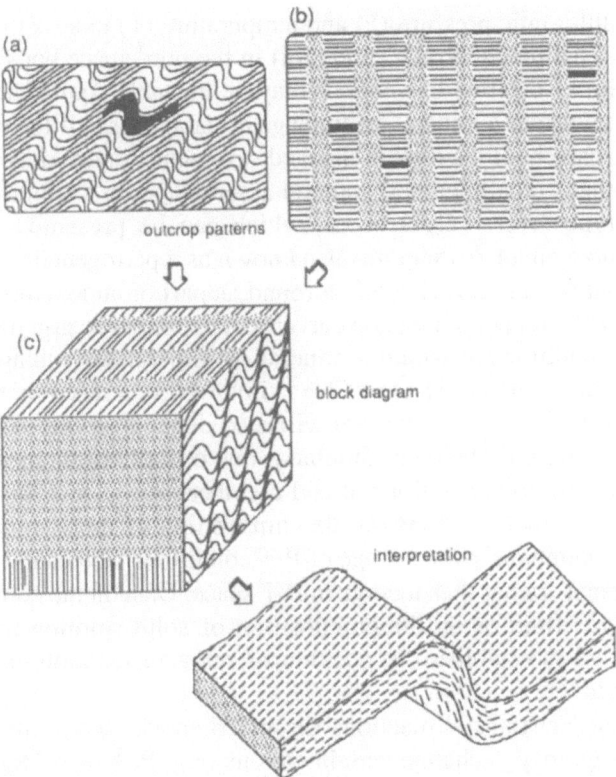

Fig. 4.46. Examples of complex geometric patterns on various outcrop surfaces of the Revenue granite, Queensland, Australia, that are explained in the text.

5 Metamorphic History of Gneiss Terrains

5.1 Introduction

The local metamorphic history is an essential topic in the study of high-grade gneiss terrains. This chapter only deals with those aspects of the metamorphic history that can be studied in the field. We will briefly outline terminology, the methods which can be applied, and the difficulties involved. For more information we refer to textbooks such as Miyashiro (1975), Winkler (1976), Mason (1981), Best (1982), Vernon (1983), Spry (1986), Yardley (1989) and to the references given in the text.

At fixed lithostatic pressure (P) and temperature (T) in a rock, and a fixed composition of the metamorphic fluid (X_{fl}) in the pore space between mineral grains, a certain association of minerals may be in stable equilibrium. Such an association is known as a mineral assemblage or paragenesis. If the physical and chemical environment of the rock is changed sufficiently, some minerals may be replaced by others, creating a new mineral assemblage. In a simple case, reactions of the type A+B = C+D occur, which can be presented as a single (univariant) curve on a P-T diagram, also known as a petrogenetic grid. On one side of the reaction curve, A+B will be found as part of an assemblage, on the other side C+D. A number of reaction curves in a petrogenetic grid may delimit a range of P-T conditions at which a mineral or mineral assemblage is stable. However, if a phase such as H_2O or CO_2 is involved in a reaction, the position of the reaction line in a P-T diagram will also depend on the partial vapour pressure of this phase in the rock. Similarly, one or more of the minerals in an assemblage may be solid solutions of end member phases, and the position of the reaction curve then depends on the composition of these minerals. Such reactions are 'continuous' over a range of P-T conditions, and can be thought of as a reaction 'band', rather than a curve in P-T space. Data on the composition of the metamorphic fluid or on the composition of solid solution minerals are therefore needed to establish the position of relevant reaction curves in the petrogenetic grid.

Besides the 'net-transfer reactions' discussed above, two or more minerals in an assemblage may exchange certain cations (e.g. Fe^{2+} and Mg^{2+} between biotite and garnet) upon a change in temperature without actual new growth of any mineral phase. Such cation exchange reactions can be used to determine the temperature at which phases were in stable equilibrium. Mineral indicators of metamorphic P or T discussed above are known as geothermobarometers. A list of geothermobarometers which can be used in high-grade rocks is given at the end of this chapter.

5.2 Metamorphic History

Metasedimentary rocks in a gneiss terrain must have travelled along a prograde path in P-T space to reach peak metamorphic conditions and then through a retrograde path back to the surface. Figure 5.1a shows a possible P-T path for a high-grade gneiss terrain. Gneiss terrains with a long history may have much more complex paths, but clockwise P-T paths of the kind shown here are relatively common (Harley, 1989; Ellis, 1987). Evidence from several segments of this path can be preserved in the mineral content of high-grade rocks, which means that not all minerals in a high-grade rock need to be part of a single peak-metamorphic mineral assemblage. During prograde metamorphism to high-grade conditions, anhydrous minerals gradually replace hydrous phases in the rock by dehydration reactions, and water is expelled along fractures or shear zones towards higher crustal levels, or absorbed by pockets of melt which may locally migrate. Near peak metamorphic conditions[1], diffusion rates are relatively high, and most relics of older assemblages are destroyed by textural and chemical equilibration of the rock. During retrograde metamorphism, the dehydration reactions mentioned above can only be reversed if there is an influx of water. In many gneiss terrains, the absense of free water under high-grade conditions means that peak metamorphic mineral assemblages are preserved, although minor corona structures may form and cation exchange may modify the chemical composition of some constituent minerals. Since diffusion rates decrease with falling temperature, the mineral composition of many high-grade gneiss terrains mainly reflects a short, specific part of the total P-T path immediately following peak metamorphic conditions (Fig. 5.1a-1; Bohlen, 1987).

---

[1] With peak metamorphic conditions we mean the maximum temperature to which the rock has been subjected (Fig. 5.1)

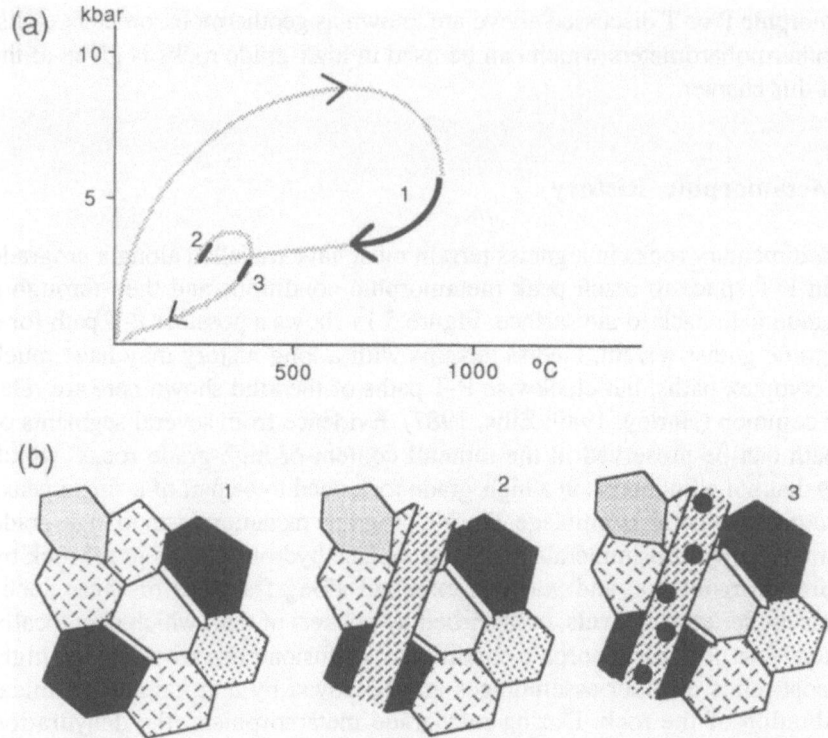

Fig. 5.1. (a) Typical P-T path reaching high-grade metamorphic conditions as a result of crustal thickening following continental collision; grey - actual path; black - parts of the path preserved in assemblages shown in (b). (1) Peak metamorphic conditions, represented by an assemblage with a granoblastic, polygonal fabric, overprinted by reaction rims during the first stage of cooling and decompression. (2) Shear zone formed during a minor second orogenic phase, preserving assemblages from point 2 on the P-T curve. (3) Growth of new phases at peak metamorphic conditions during the second orogenic phase at point 3 on the P-T curve.

5.3 Fabric Evidence for Metamorphic History

It is obviously of crucial importance to determine where on the P-T path different components of the mineral fabric of a rock formed during the meta-morphic history. Most of this work must be done in thin section, but it is important to obtain as much information as possible in the field. Data on the relationship of stable mineral assemblages with deformation phases and intrusive events, for instance, are difficult to determine in thin section because of the coarse-grained nature of most high-grade rocks.

Mineral assemblages that formed during peak metamorphic conditions are usually dominant in high-grade gneiss terrains. They are characterised by un-zoned crystals which are in contact without the presence of reaction rims and which commonly have a granoblastic polygonal shape (Section 3.3.1; Figs. 3.6a; 5.1b-1). Coarse-grained polygonal assemblages were probably stable during the metamorphic peak, even though mineral pairs are separated by coronas.

Relics of the prograde metamorphic path may be found in lenses of low strain, such as boudins. However, it is necessary to be well aware of the structural history of an area before interpreting the relative age of assemblages. Relics of older assemblages, even hydrous ones, may also be enclosed by, and preserved in other minerals. For example, kyanite inclusions inside large garnets in a sillimanite-bearing high-grade gneiss may indicate that the prograde P-T path crossed the kyanite-sillimanite reaction curve.

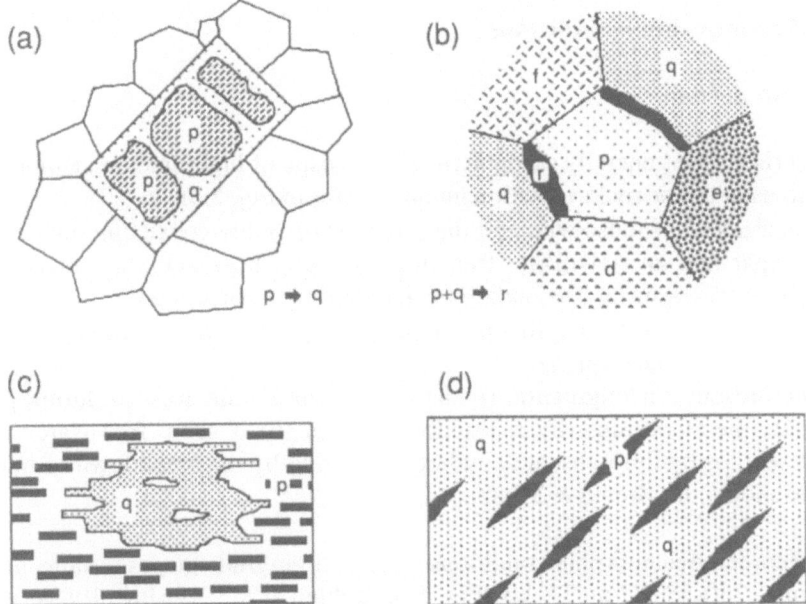

Fig. 5.2. Examples of the relative dating of metamorphic events. (a) Pseudomorphic replacement of phase (p) by phase (q) - parallelism of the cleavage in different fragments of (p), and the margin of the pseudomorph indicate that (p) is the older phase. (b) The presence of reaction rims of a mineral (or polymineralic symplectite) (r) between (p) and (q) indicates the instability of p+q. (c) Pseudomorphic replacement of minerals (p) in a foliation indicate that (q) postdates the deformation that formed the foliation. (d) Exsolution lamellae of phase (p) in phase (q).

Fabric elements that formed during retrogression can be recognised by several incomplete replacement structures. The most common ones are (Fig. 5.2):

(1) Pseudomorphic replacement of a mineral grain **p** along the rims and along cleavage planes, cracks or kinks by a younger phase **q** or symplectitic growth of **q** and **r** (Fig. 5.2b). Remnants of the older phase can be recognised by the common orientation of their cleavage (Fig. 5.2a).

(2) Reaction rims of a younger phase **r** or symplectitic rims of more than one phase around mineral grains of an older phase **p** where they are in contact with **q** (Figs. 5.1b-1; 5.2b; 5.5). In some cases, isolated new grains of the new minerals form instead of complete rims.

(3) Grains of a younger phase **q** mimicking a preferred orientation (foliation or lineation) of grains of an older phase **p** (Fig. 5.2c). This also provides evidence for the relative age of the reaction with respect to deformation.

(4) Exsolution lamellae of mineral **p** in **q**, which form by reactions of the type **q** (rich in p) = **q** (poor in p)+**p,** when the chemical composition of **q** becomes unstable with changing metamorphic conditions (Fig. 5.2d). An example is perthitic albite lamellae in K-feldspar.

5.4 Metamorphic Conditions

5.4.1 Introduction

Three criteria are generally recognised as diagnostic of high-grade metamorphic conditions in pelitic rocks (reaction numbers refer to Fig. 5.3):

(1) The absence of muscovite and the presence of stably coexisting quartz, K-feldspar and an Al-silicate (reaction 1). Note, however, that muscovite generally appears as a secondary mineral during retrogression.

(2) The presence of orthopyroxene (hypersthene), or of the assemblage clino-pyroxene+ garnet+ quartz.

(3) The presence of migmatitic (melt) veins. These veins formed during pro-grade metamorphism by a reaction such as (6) if some free water was present, or by fluid-absent dehydration melting by the breakdown of micas (reaction 5; Le Breton & Thompson, 1988).

Recorded metamorphic conditions in most high-grade gneiss terrains lie in the range 750-850°C at 5-8 kb (Fig. 5.3; Bohlen, 1987; Harley, 1989), well within the field of 'wet' granite melting (indicated by reaction 6 in Fig. 5.3). Nevertheless, high-grade gneisses usually contain only a small proportion of locally formed melt. This can be attributed to the low H_2O activity which is characteristic of high-grade metamorphic terrains (Newton, 1988; Valley, 1988). A possible explanation for this low H_2O activity is infiltration of CO_2 rich fluids, possibly from the mantle, which promoted dehydration (Touret, 1971; Hansen et al., 1987). A great deal of attention has recently been focused on the so-called 'in situ granulitisation' phenomenon in lower crustal rocks, described from southern India, Sri Lanka, Madagascar and Antarctica. In these terrains, patches of charnockite or orthopyroxene-bearing quartzo-feldspathic assemblages developed on a cm to m scale, and give the rock a spotted appearance (locally referred to as "measles rock"). In some cases it is thought that formation of high-grade assemblages did not require any increase in temperature and/or pressure but resulted from an influx of CO_2 while water was purged from the system, thus

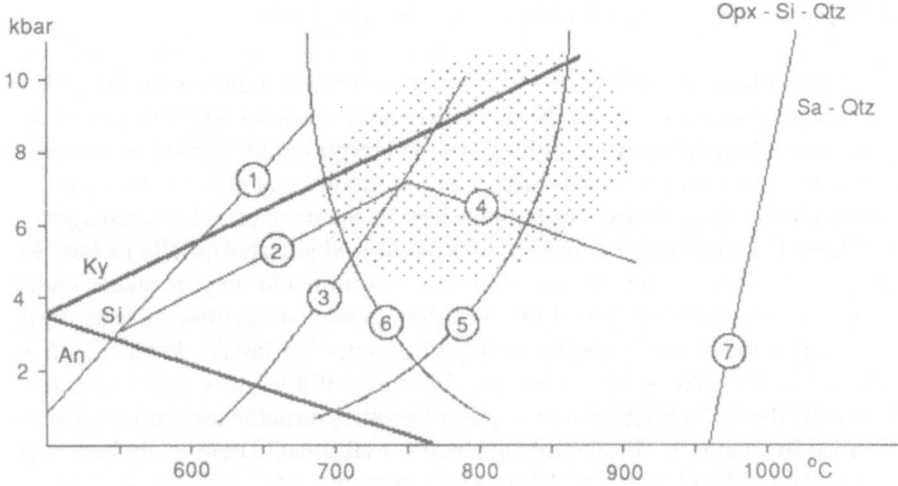

Fig. 5.3. Petrogenetic P-T grid showing common reactions in pelitic rocks at high-grade metamorphic conditions. Grey field: range of most commonly encountered conditions in high-grade metamorphic rocks, after Harley (1989). Approximate conditions where orthopyroxene+sillimanite+quartz and sapphirine+quartz can be found are indicated (Harley, 1989). Bold lines represent univariant reaction curves between stability fields of the Al-silicates (Holdaway, 1971). Thin curves represent continuous reactions and can only provide a general impression of the distribution of phase changes during high-grade metamorphism. The reactions are[1]:

1. Mu+Q=Kf+Als+V; (X_{H2O} = 0.4; Chatterjee & Johannes, 1974)
2. Bi+Als+Q=Cd+Kf+V; (X_{H2O} = 0.4 and X_{Fe} = 0.2 in cordierite; Holdaway & Lee, 1977)
3. Bi+Als+Q=Ga+Cd+Kf+V; (X_{H2O} = 0.4 and X_{Fe} = 0.2 in cordierite; Holdaway & Lee, 1977)
4. Cd=Als+Ga+Q+V; (X_{H2O} = 0.4 and X_{Fe} = 0.2 in cordierite; Holdaway & Lee, 1977)
5. Ph+Q = Kf+En+melt (Peterson & Newton, 1989)
6. Plag+Q+Kf+V = melt (X_{H2O} = 0.4; Holdaway & Lee, 1977);
7. Plag+Q+Kf = melt (in absence of biotite; Huang & Wyllie, 1975).

effectively dehydrating the rock assemblage (Newton et al., 1980; Newton and Hansen, 1983). This CO_2 streaming hypothesis is quite popular at present among some geologists, but the origin of the CO_2 is unclear. It should also be noted that some stable isotope data are incompatible with this hypothesis (Lamb and Valley, 1984; Fiorentini et al., 1990).

Alternative explanations for the scarcity of locally formed melt material in high-grade gneisses is 'desiccation' of a terrain by shallow level intrusives prior to high-grade metamorphism (McLelland and Hussain, 1986), or dissolution of free H_2O during prograde metamorphism in partial melt fractions which have been removed from the rock (Powell, 1983).

---

[1] The following abbreviations are used here: Als -Al-silicate; An -andalusite; Amp -amphibole; Bi -biotite; Cd -cordierite; Cpx -clinopyroxene; En -enstatite; Ga -garnet; Ilm -Ilmenite; Kf -K-feldspar; Ky -kyanite; Mu -muscovite; Opx -orthopyroxene; Plag -plagioclase; Ph -phlogopite; Q -quartz; Rut -rutile; Sa -sapphirine; Si -sillimanite; Sph -sphene; V -vapour.

5.4.2 Mineral Assemblages in High-Grade Metamorphism

It is impossible to accurately determine metamorphic conditions in high-grade rocks without accurate data on the chemical composition of individual minerals in an assemblage. However, high-grade metamorphic conditions can be distinguished and roughly subdivided using mineral assemblages that can be observed in the field. Below, we outline some of the most useful assemblages.

Mineral assemblages in gneisses with Fe-Mg-Al silicates usually include K-feldspar, plagioclase and quartz. Biotite and hornblende may be stable under high-grade metamorphic conditions even though they are hydrous phases. With in-creasing temperature in pelitic rocks, muscovite first breaks down (Fig. 5.3; reaction 1), followed by the assemblage biotite + sillimanite + quartz (reaction 3). Finally, the assemblage biotite + quartz becomes unstable and orthopyroxene is formed (reaction 5). The assemblage biotite + sillimanite may easily be recognised in the field and indicates the lower temperature range of high-grade metamorphism. Above the solidus, mica-breakdown reactions produce small volumes of melt in the form of migmatite veins which may contain garnet or cordierite (reaction 3), or orthopyroxene (reaction 5). Any Fe-Mg-Al silicates present in melt pockets may therefore be useful indicators of metamorphic grade.

The cordierite-orthopyroxene assemblage in rocks of lower to intermediate bulk (Mg/Mg+Fe) content is often indicative of very low pressure conditions, up to 3 kb (Harris & Holland, 1984; Vielzeuf & Hollaway, 1988). Garnet gradually replaces cordierite towards higher pressure conditions. The stable co-existence of K-feldspar + sillimanite indicates intermediate pressure conditions, K-feldspar + kyanite high pressures.

Although most gneiss terrains reached peak metamorphic conditions at temperatures below 850°C, some 'very high-grade' terrains may have reached temperatures of over 1000°C. They are characterised by the assemblages sapphirine + quartz (very high temperature) or hypersthene + sillimanite + quartz (very high pressure and temperature; Fig. 5.3). Fe-rich spinel + quartz is also typical of very high temperatures under low to medium pressure conditions.

In mafic rocks, clinopyroxene, hypersthene, almandine, plagioclase and quartz may coexist under high-grade conditions. Biotite, hornblende and ilmenite may also be present. Towards high pressures, orthopyroxene is progressively replaced by garnet; Under high pressure conditions, orthopyroxene is lacking in plagioclase bearing clinopyroxene metabasites. Ultramafic rocks may also enable a rough assessment of metamorphic conditions in the field. The presence of the assemblages [anthophyllite or cummingtonite]+[olivine or enstatite] or [anthophyllite or cummingtonite]+[enstatite or quartz] suggests high-grade conditions.

Calc-silicate rocks can be good indicators of metamorphic grade, but they are not easy to interpret in detail in the field, since P-T stability fields of the assem-blages strongly depend on the composition of the metamorphic fluid. Wollas-tonite with or without calcite, and periclase or spinel coexisting with calcite are indicative of high metamorphic grade. Forsterite, scapolite, diopside and grossular may also be present.

5.4.3 Mineral Assemblages in High- to Medium-Grade Retrogression

Retrograde P-T paths from peak metamorphic conditions may have a complex shape, and two end-member types are generally distinguished (Harley, 1989): cooling with little decompression (isobaric cooling IBC); and rapid decompression with little cooling (isothermal decompression ITD). The first path is thought to be characteristic of gneiss terrains which are resident at depth in a crust of normal thickness and which gradually cool there over a long period of time. The thermal perturbation needed to create and maintain the high-grade conditions may have been caused by heat supplied during magmatic underplating or voluminous intrusions (Chapter 7); the second type is thought to be more typical of uplift and erosion of continental crust that was thickened by collision. Certain mineral reaction fabrics, which may be visible in the field in coarse-grained rocks, are considered to be diagnostic of IBC or ITD paths; however, microstructural and microprobe analyses of constituent phases are always needed to confirm this (Harley, 1989).

IBC paths can be recognised in high-grade metabasic rocks by garnet rims or Ga-Q and Ga-Cpx-Q symplectites on orthopyroxene-plagioclase contacts (Fig. 5.4a; Ellis & Green, 1985; Harley, 1989); by clinopyroxene rims on orthopyroxene; or by beads or lamellae of garnet in aluminous pyroxene (Fig. 5.4b). In pelitic rocks, rims of garnet enclose older garnet or orthopyroxene.

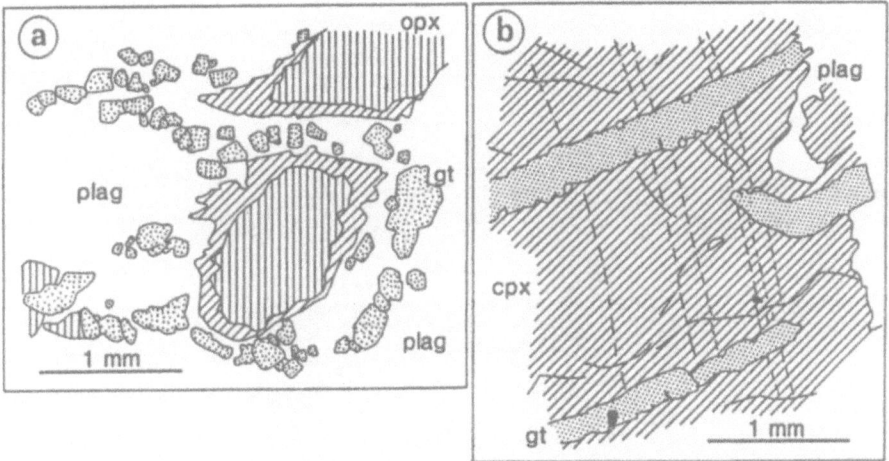

Fig. 5.4. Some examples of textures associated with IBC paths, after Harley (1989). (a) Bead-like or sugary garnet developed on plagioclase grain boundaries, and clinopyroxene occurring as mantles on orthopyroxene in a metabasite, Hydrographer Island, Enderby Land. (b) Garnet occurring as lamellae or rods along kink-band boundaries in deformed Fe-rich aluminous clinopyroxene. Pyroxenite, Enderby Land.

ITD paths in high-grade metabasic rocks are characterised by the breakdown of older garnet; characteristic are orthopyroxene-plagioclase symplectites on garnet and orthopyroxene rims on clinopyroxene (Fig. 5.5a,b). In high-grade pelitic and felsic rocks, plagioclase rims occur on garnet or garnet-orthopyroxene contacts (Harley, 1988). Also, Opx-Cd symplectites may form between garnet and quartz (Fig. 5.5c). Small pockets of melt may also form under these conditions if the P-T path crosses dehydration curves such as reaction (5) in Fig. 5.3.

5.4.4 Mineral Assemblages in Low-Grade Retrogression

In most high-grade metamorphic terrains, mineral assemblages can be found that formed under low-grade conditions. Minerals to be expected in pelites include biotite, epidote, sphene, albite, chlorite, and white mica. In mafic rocks, biotite, actinolite, epidote, sphene, albite, chlorite, prehnite, pumpellyite and carbonate occur. In calc-silicates, tremolite and talc appear; in ultramafic rocks talc, chlorite and serpentine are formed.

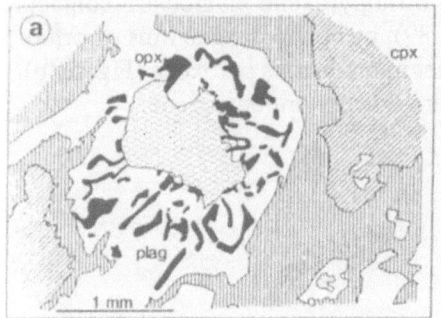
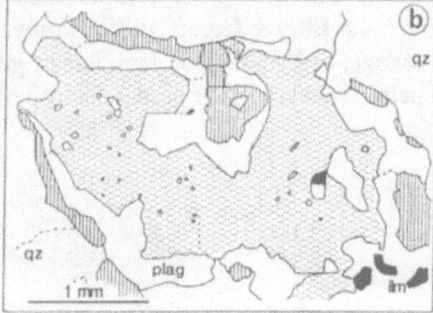
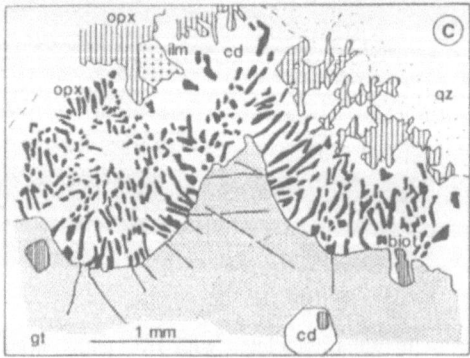

Fig. 5.5. Some examples of textures associated with ITD paths, after Harley (1989). (a) Opx-Plag symplectite on resorbed garnet, and orthopyroxene mantling clinopyroxene in a metabasite. Rauer Islands, Antarctica; (b) Plag-Opx moat and partial rim between garnet and quartz. The plagioclase is more calcic than primary plagioclase from the same sample. Felsic high-grade rock, Rauer Islands. Antarctica; (c) Opx-Cd symplectitic corona adjacent to and resorbing garnet, and coarser orthopyroxene adjacent to quartz in initial Ga-Opx-Q metapelite from the Sharyzhalgay Complex, Lake Baikal.

5.5 Sites to Study Retrogression Fabrics

Replacement structures typical of retrogression (Fig. 5.2) are preferentially developed in the following sites:

(1) Sites of strong deformation such as boudin necks or shear zones (Fig. 5.1b.2), with the same bulk composition as the host rock. In such sites, an older mineral assemblage is ductilely deformed, which means that the lattice of the individual crystals is affected by crystalplastic deformation, or microcracks. If a shear zone was active during retrogression, the material in the shear zone would attain a new fabric and new compositional equilibrium due to enhanced diffusion in the damaged crystal lattices, while the assemblage in the wall rock would be preserved. Shear zones also act as important channels for fluids. The effects of retrograde reactions which involve the growth of hydrous mineral phases at the expense of water-free minerals are therefore commonly observed in shear zones. In some cases, several generations of crosscutting shear zones may occur in an area, each with a distinct mineral assemblage (Fig. 5.6).

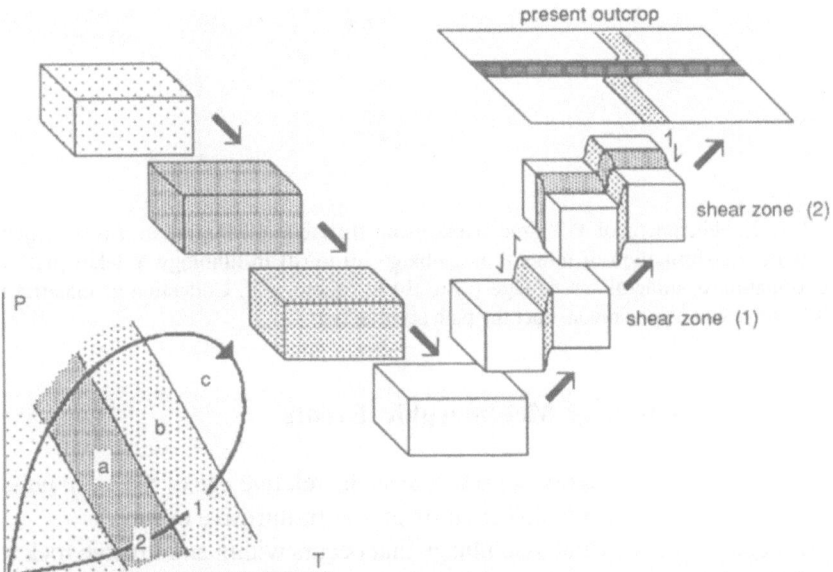

Fig. 5.6. Establishment of P-T-time paths using transecting shear zones; mineral assemblages formed during prograde metamorphism are not preserved, but during uplift two phases of shear zone development preserve two assemblages of the retrograde path.

(2) Melt pockets which solidify during retrograde metamorphism. These produce free water which can hydrate mineral assemblages in the wall rock. Biotite + sillimanite rims which are commonly observed around such melt pockets may have formed in this way (Section 4.7.5).

(3) Domains with a different bulk chemistry from the host rock, such as primary compositional variations, intrusions, partial melt veins and shear zones in which fluid fluxing has changed the composition. Such domains may contain mineral assemblages which are younger or older than those in the host rock. Comparison of these assemblages and the bulk chemistry of the domains can help to establish the local metamorphic history (Fig. 5.7).

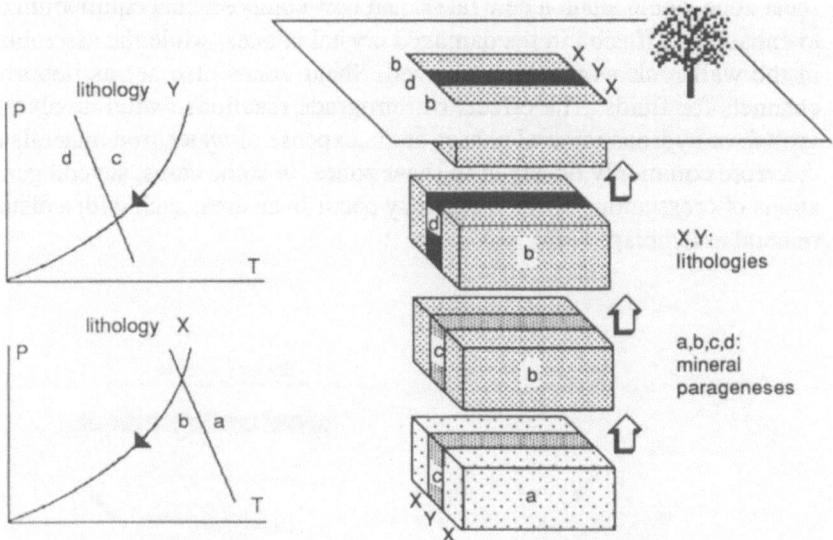

Fig. 5.7. Establishment of P-T-time paths using different lithologies in a rock sequence; retrograde transformation of mineral assemblage (c) to (d) in lithology Y takes place after transformation of mineral assemblage (a) to (b) in lithology X. Collection of material from both X and Y can help to reconstruct the path accurately.

5.6 Relative Dating of Metamorphic Events

The following criteria can be used to assess the relative age of a certain point on the P-T path with respect to deformation phases or intrusive events:

(1) A metamorphic mineral assemblage that occurs within an intrusive rock must be synchronous with or postdate the emplacement of that intrusion. The same argument applies if several intrusive phases are present with different parageneses. The same assemblage in several intrusions, however, may mean that the older ones have been re-equilibrated. Because of the high water content of most melts, mineral assemblages in melt veins or intrusions may correspond to retrograde conditions rather than the actual conditions at which the veins formed.

(2) Mineral assemblages which only occur within domains of high strain such as shear zones, fold closures and boudin necks must be at least as young or

younger than that structure, provided that the composition of the deformed domain is the same as that of the wall rock. If they differ, caution is needed since the observed effect may be due to different rock compositions, or the paragenesis in the structure may actually predate that in the host rock if the latter has been re-equilibrated.

(3) If minerals defining a foliation or lineation have recrystallised during or after the development of the structure, they must have re-equilibrated to new conditions at that stage. These minerals are then dated with respect to a single phase of deformation. Mineral lineations are particularly important, since they imply new mineral growth during deformation. Always consider the possibility, however, that mineral grains have only rotated to form a fabric, without recrystallisation and re-equilibration.

(4) The study of inclusion trails in porphyroblasts is particularly useful to assess the relative ages of mineral growth and deformation (Yardley, 1989; Vernon, 1978; 1983); in coarse-grained rocks these inclusion trails can be observed in the field.

5.7 Isograd Patterns

Within most high-grade gneiss terrains, rocks with identical composition may contain different assemblages. Within a single rock unit, the boundary between such assemblages can be mapped and is known as an isograd. If several isograds can be mapped, the result is an isograd pattern. In a few cases, even the shape of three-dimensional isograd surfaces can be established. Isograd patterns are a useful tool for determining the tectonic history of a gneiss terrain.

If isograd surfaces are planar and near-horizontal, they may reflect a simple metamorphic gradient of increasing P-T conditions with depth. In many cases, gneiss terrains became slightly tilted during uplift, and isograds can then be used to determine the direction and amount of tilt. If isograd surfaces are dome-shaped, they may reflect intrusion or solid state diapir development (Schwerdtner, 1980, 1982). Large scale folding, thrusting or extension may lead to complex isograd patterns. Careful mapping and sampling is required to determine which mechanism is responsible. The following potential problems should, however, be considered when isograd patterns are analysed:

(1) Isograd maps only have significance if they are constructed for a single rock type.

(2) In addition to regional variations in P and T, the presence and composition of a volatile phase has an influence on some reactions in metamorphic rocks and therefore on the position of isograds.

(3) Shear zones which postdate the establishment of an isograd pattern may cause so much displacement that isograds coincide with late shear zones in outcrop. Mapping in this case reveals the geometry of late major shear zones, not the original isograd pattern (Fig. 5.8). Even if this is not the case, shear zones can change the distance between isograds. Geothermal gradients

may be grossly over- or underestimated in such cases if shear zones have not been recognised.

(4) Dome-shaped isograd patterns may be original features, formed during the metamorphic event, but polyphase fold interference on a large scale may produce the same pattern. If dome-shaped isograd patterns are encountered, a structural analysis should be made of the area.

(5) Different isograds in a terrain can be of different age. The distance between isograds can therefore rarely be used with confidence to calculate field P-T gradients. Establishment of the relative age of mineral assemblages on both sides of an isograd and phases of deformation or intrusion can help to solve this problem.

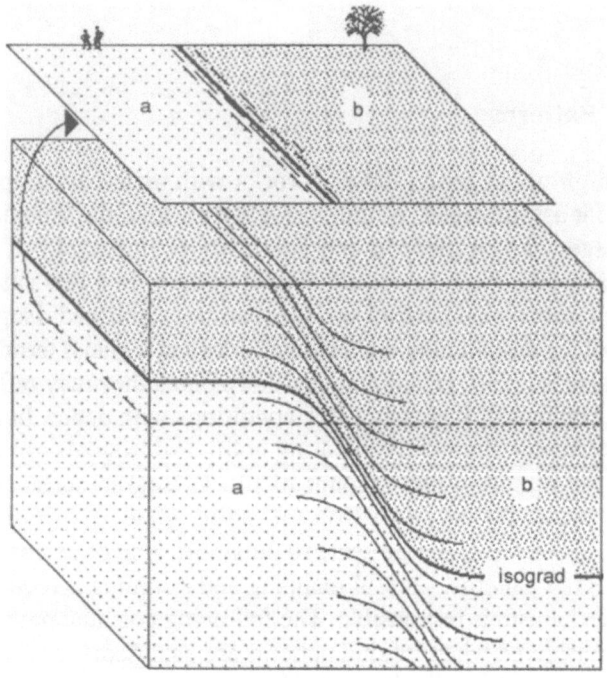

Fig. 5.8. Transection of an isograd by a shear zone can result in coincidence of the isograd and the shear zone in outcrop.

5.8 Geothermobarometry

A list of useful geothermobarometers in high-grade gneiss terrains is given below. Equilibration used for geobarometers are mostly of the net-transfer reaction type, while most geothermometers are based on cation-exchange equilibria. For further details see Essene (1982), Vielzeuf & Boivin (1984), Lagache (1984), Harley (1989) and Hensen (1987).

Geobarometers:

Ga-Rut-Ilm-Als-Q	Bohlen et al. (1983)
Ga-Plag-Opx-Q	Newton & Perkins (1982), Bohlen et al. (1983)
Ga-Plag-Q-Cpx	Perkins & Newton (1981)
Ga-Si/Ky-Q-Plag	Newton & Hasilton (1981), Anovitz & Essene (1987), Koziol and Newton (1988)
Opx-Ga	Wood (1974), Harley & Green (1982), Harley (1984a)
Ga-Cd-Si-Q	Hensen & Green (1973), Aranovich & Podlesskii (1983)
Ga-Cd-Opx-Q	Hensen & Green, (1973), Aranovich & Podlesskii (1983)

Geothermometers:

Ga-Cpx	Ellis & Green (1979)
Ga-Opx	Harley (1984b), Lee and Ganguly (1988)
Ga-Bi	Ferry & Spear (1978)
Ga-Cd	Thompson (1976), Perchuk & Lavrent'eva (1983)
Ga-Amp	Graham & Powell (1984)
Opx-Cpx	Lindsley (1983)

6 Geochemistry, Isotope Geochemistry and Geochronology: Application to Field Studies

6.1 Introduction

Structural studies make it possible to reconstruct the metamorphic and deformation history of a rock body but do not provide information on the absolute timing of these processes. Likewise, field observations and structural work alone may not unambiguously identify the protoliths of many high-grade gneisses or the compositional changes associated with migmatisation, anatexis and the general action of fluid and/or vapour phases in a rock. Geochemical research can answer many of these problems but, as most of this research is laboratory-oriented, we limit ourselves to some general outlines of the possible lines of research, so that these can be borne in mind during fieldwork.

6.2 Geochemistry

Geochemical data indicate that some high-grade rocks are severely depleted in the large incompatible (LIL) elements such as U, Th, K, Rb and Ba (e.g. Fig. 6.1). These elements appear to have been removed from high-grade rocks while they were resident in the lower crust, probably during high-grade metamorphism, and through the action of fluid, vapour or melt phases, were transported to, and concentrated in, the upper crust. Geochemistry may be able to identify such depletions and the processes associated with them. In some cases it can be recognised that rocks in high-grade terrains are 'restites' and that a granitic melt fraction has been removed from them (e.g. Clemens & Vielzeuf, 1987); in other cases it may be more likely that a vapour phase such as CO_2 mobilized LIL elements and that the remaining rock is not a restite (Rudnik & Presper, 1990). Both processes result in rather unusual chemical compositions in some lower crustal rocks (e.g. Schenk, 1984), and the petrogenetic changes that occurred can be identified with the help of geochemical data. For instance, restites generally exhibit a positive Eu-anomaly in their Rare Earth Element (REE) patterns because, under conditions of reduced O_2 activity, Eu is concentrated in the plagioclase lattice and stays behind during partial melting and migration of a granite liquid.

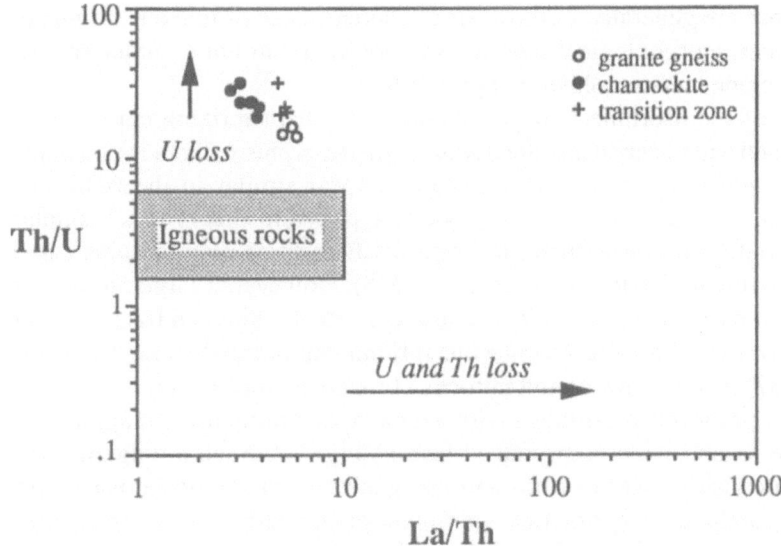

Fig. 6.1. Plot of La/Th versus Th/U. Shaded box shows field of upper crustal igneous rock types (from Rudnik and Presper, 1990). Data points above this field represent granite gneisses and charnockites from Sri Lanka and display marked U-loss (data from Milisenda, unpubl.).

Other high-grade gneiss terrains, however, do not exhibit this depletion pattern, and there was apparently little or no change in the bulk rock chemistry during development of high-grade mineral assemblages. In these cases even 'flushing' large amounts of CO_2 through the crust (Chapter 5) did not result in marked changes in geochemistry (e.g. Stähle et al., 1988). In a compilation of geochemical data from high-grade rocks of various ages throughout the world, Rudnik & Presper (1990) observed that Cs and U were almost always depleted relative to upper crustal rocks whereas Rb and Th were variably depleted, and other elements show no diagnostic behaviour at all.

The identification of the protolith is one of the most important steps in the analysis of high-grade terrains (Leake, 1964). This is particularly difficult where intense deformation has obliterated all original structures and textures and where a pervasive layering has resulted from a combination of structural and metamorphic processes (Chapter 4). Some of the most difficult problems, for example, are to recognize whether a grey, layered gneiss was derived from a greywacke-type sedimentary rock or a granitoid rock, and to distinguish between clastic sedimentary rock, tuff or lava in a supracrustal succession. In these cases geochemistry as well as heavy mineral analyses may help. For example, the chemical data, in combination with petrology, may provide patterns that enable distinction between sedimentary and igneous rocks. Thus, pelitic sedimentary rocks have a characteristic pattern that does not change much with increasing metamorphism (Nesbitt & Young, 1984). The occurrence of Al-rich minerals such as sillimanite, corundum, sapphirine, kornerupine and osumilite are generally diagnostic of sedimentary rocks (Grew, 1984; Schreyer, 1985), and

110

detrital zircons are generally well rounded, whereas those of igneous origin are euhedral. These morphological features are modified, but not completely lost, during high-grade metamorphism (Hoppe, 1963).

In spite of the usefulness of geochemistry in characterizing certain rock types and identifying chemical changes during metamorphism, there is no unique way of distinguishing some rock types that appear similar in the field. For instance, the major geochemistry of greywacke-type sedimentary rocks is similar to that of tonalites and trondhjemites, only REE-patterns may in some cases permit a distinction (Taylor & McLennan, 1985). However, a large number of analyses often makes it possible to recognise consistent major and trace element variations that can be ascribed to igneous differentiation trends (e.g. Tarney & Weaver, 1987) that are only found in rocks of magmatic origin. Werner (1987) has recently proposed a simple major element discrimination diagram for intermediate to acid rocks, reproduced here as Fig. 6.2. Such a diagram may help to distinguish between ortho- and paragneisses but should be used with caution and, perhaps, in conjuction with other parameters such as the Niggli-values c against al-alk (Burri, 1959) and petrological criteria.

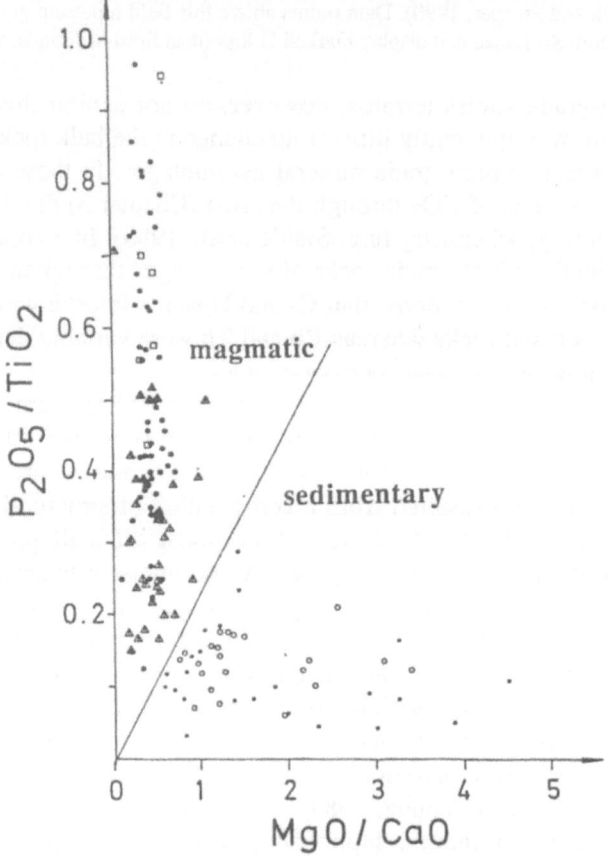

Fig.6.2. Major element discrimination diagram distinguishing ortho- and paragneisses in high-grade metamorphic terrains. Data and diagram from Werner (1987).

6.3 Isotope Geochemistry

Isotope geochemistry, in particular the Rb-Sr, Sm-Nd and U-Th-Pb whole-rock and mineral systems as well as O-isotopes, can provide important constraints on the source, evolution, and age of high-grade rocks. However, it should be noted that all these systems are affected to some extent by mineralogical and geochemical changes in the course of metamorphic recrystallisation and concurrent fluid activity so that interpretation of the isotopic data may be ambiguous (see discussion in Moorbath & Taylor, 1986).

Oxygen isotopes (the isotopic ratio $^{18}O/^{16}O$) are often useful in distinguishing between ortho- and paragneisses in high-grade terrains (e.g. Taylor, 1980). This is because the $^{18}O/^{16}O$ ratio (expressed as $\delta^{18}O‰$) is generally high in sedimentary rocks (12-30 or more $\delta^{18}O‰$), while it is low in igneous rocks derived from melting of mantle-derived protoliths (5-10 $\delta^{18}O‰$) and intermediate in rocks resulting from intracrustal melting (7-13 $\delta^{18}O‰$). These values do not change much during high-grade metamorphism, and it has been shown recently that marked variations in O-isotope values found in some high-grade rocks argue against wholesale "flushing" of CO_2 through the rock (Fiorentini et al., 1990).

The use of radiogenic isotopes in the investigation of high-grade terrains has certain limitations that result from the fact that high-grade metamorphism generally changes the original isotopic ratios and may also affect the parent-daughter ratios (particularly Rb/Sr and U/Pb) so that the pre-metamorphic history is lost. The rock is said to be isotopically "reset". This has severe implications for both geochronology (Section 6.4) and petrogenetic interpretation. For example, calculation of a 'model age' (see below) in the Rb-Sr and Sm-Nd systems largely depends on the parent-daughter isotopic ratio (e.g. $^{87}Rb/^{86}Sr$ and $^{147}Sm/^{144}Nd$, see Faure, 1986). If the isotopic ratios have decreased during metamorphism, e.g. by preferential removal of Rb and Sm from the rock system at sample scale, the apparent model ages calculated may be too low (Fig. 6.3). Until recently, it was generally believed that the Sm-Nd system is insensitive to high-grade metamorphic recrystallisation, but a few cases have been reported where this is apparently not the case. This applies to high-grade rocks in Enderby Land, Antarctica (DePaolo et al., 1982; McCulloch & Black, 1984; Black & McCulloch, 1987) and to the Lewisian complex, Scotland (Whitehouse, 1988). The Sm-Nd and Pb-Pb systems are particularly useful in identifying the effect of crustal contamination in magmatic rocks and in reconstructing crustal growth processes (e.g. DePaolo, 1981; Moorbath & Taylor, 1986).

It has become fashionable to calculate so-called model ages for crustal rocks, using the Sm-Nd whole-rock system (McCulloch & Wasserburg, 1978). Such model ages or, as they have been called more correctly, crust-formation or mantle-extraction ages, indicate the approximate time a given rock or its precursor separated from the mantle during a mantle-crust differentiation process.

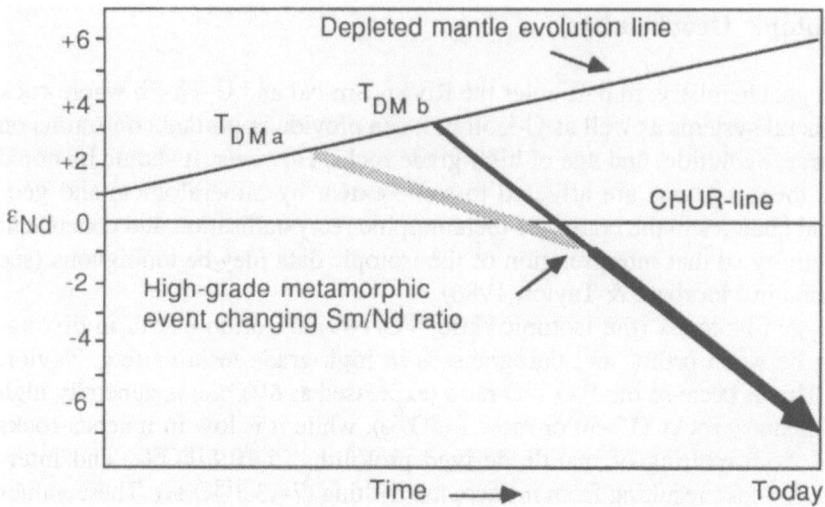

Fig. 6.3. Sm-Nd evolution diagram showing how model ages (mantle extraction ages) are determined and why the model age depends heavily on the Sm-Nd ratio of a given rock (modified from DePaolo, 1981). In the case above, the rock has experienced high-grade metamorphism changing the original Sm/Nd ratio. The model age calculated on the basis of the Sm/Nd ratio measured today (T_{DMb}) is too low. The correct model age (T_{DMa}) depends on the original Sm/Nd ratio of the rock prior to metamorphism and is usually difficult to determine.

In the case of juvenile magmatic rocks and their metamorphic derivatives (volcanics and intrusives derived from melting of a mantle-derived precursor) this model age may be close to the actual emplacement age, whereas clastic sedimentary rocks and crustally derived magmatic rocks (S-type granites and many felsic volcanics) largely inherit mixed isotopic source signatures and hence yield mean crustal residence ages, i.e. the mean time that the precursor(s) of the analysed rock has/have been in the continental crust. Since several assumptions on the isotopic composition and evolution of the upper mantle must be made, and since we know that the upper mantle is, and was, isotopically heterogeneous with respect to Sm-Nd, the model "ages" are only approximations of the above events and may have inherent errors of up to several hundred million years (Arndt & Goldstein, 1987). A model age should therefore not be confused with a geochronologic age.

A good example of the use of the Sm-Nd system to separate crustal units on the basis of model ages and isotopic systematics is provided by data from Sri Lanka (Milisenda et al., 1988). Here it was possible to separate an old crustal province with model ages as old as Archaean (the Highland and Southwest Group supracrustal rocks and related S-type granites) from a juvenile terrain (the Vijayan Complex), consisting of I-type granites and interpreted as part of a younger island arc system (Fig. 6.4).

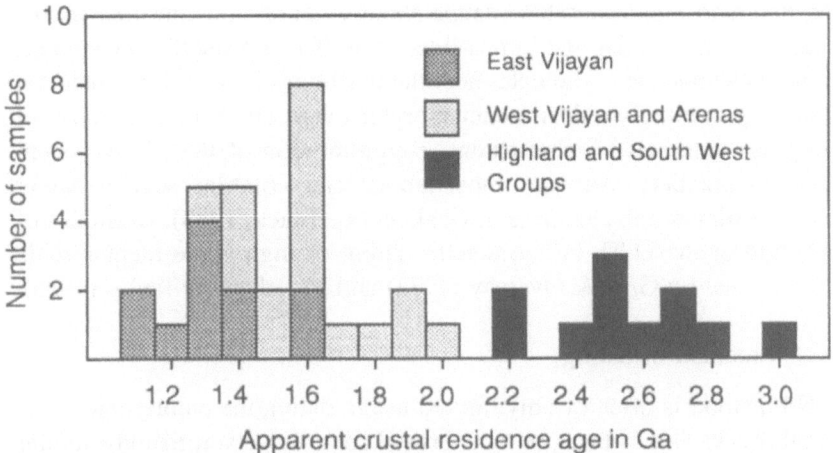

Fig. 6.4. Histogram showing differences in Nd model ages between Vijayan gneisses and rocks of the Highland and Southwest Groups, Sri Lanka (from Milisenda et al., 1988).

6.4 Geochronology

6.4.1 Introduction

Another important use of isotopic systematics is in dating rocks and minerals. Whole-rock samples are usually preferred for dating using the Rb-Sr, Sm-Nd and Pb-Pb systems, while mineral ages are determined by the K-Ar and $^{39}Ar/^{40}Ar$ (muscovite, biotite, hornblende), and U-Pb (zircon, monazite, sphene, garnet, apatite) methods. If information on metamorphic ages is required, metamorphic minerals such as garnet, clinopyroxene, orthopyroxene, hornblende, plagioclase, K-feldspar and biotite, usually together with the whole rock, can be dated by the Sm-Nd, Rb-Sr, and $^{40}Ar/^{39}Ar$ methods, and in the case of garnet, also by the U-Pb method. These techniques require clean separation of the above minerals in the laboratory, and this is usually a laborious process. Since the different isotopic systems have different s (the temperature below which the system becomes blocked because isotopic exchange is no longer possible, Dodson, 1979), they can be used to date different geologic events. For example, the K-Ar system has one of the lowest closure temperatures, and in metamorphic rocks usually dates times of crustal uplift. Thus, it has been possible to reconstruct the uplift histories of the Alps and Himalayas and relate them to collision processes.

The Rb-Sr system is usually strongly affected by high-grade metamorphism and may be completely or partly reset (e.g. Peucat, 1986; Black, 1988). In some cases, this can be overcome by collecting very large samples (20 kg or more) because examples are known in which isotopic mobility during regional meta-morphism was limited to a scale of only a few centimetres. However, in general, ages produced by the Rb-Sr method must be regarded with caution, and many published data show considerable scatter about a regression line that makes calculation of a meaningful age impossible. Even if a statistically meaningful 'age' can be calculated (i.e. the data meet the isochron criteria), it is uncertain whether this 'age' actually reflects a metamorphic event, since it is generally not known at what time and to what extent re-equilibration of the Rb-Sr system took place. In practice, therefore, most laboratories combine several dating techniques to estimate the age of geological events (Black, 1988). Examples of the application of the U-Th-Pb system for dating of high-grade metamorphic rocks are provided by Gray & Oversby (1972) and Moorbath & Taylor (1986).

6.4.2 Sm-Nd method of dating

The Sm-Nd method is often of only limited use in dating the emplacement age of granitoid rocks. This is because most crustal rocks have surprisingly similar Sm/Nd ratios so that the spread in isotopic ratios is insufficient to calculate accurate ages by the isochron method. One can overcome this problem, however, by combining mineral and whole-rock isotopic data because considerable fractionation of the Sm-Nd system takes place in minerals (e.g. orthopyroxene and plagioclase have low Sm/Nd ratios while garnet, clinopyroxene and zircon have high ratios). De Paolo (1988) provides an excellent introduction into the isotopic systematics of the Sm-Nd system and its application to geochronology.

6.4.3 U-Pb dating of zircons

The most reliable method to date the emplacement age of an igneous rock and its subsequent high-grade metamorphic history is provided by the U-Pb isotopic system in zircon. This mineral incorporates U into its lattice during crystallisation and, although some radiogenic Pb may diffuse out of the crystal during medium to high-grade metamorphism and again during weathering and/or groundwater leaching, the original crystallisation age is hardly ever lost and can be determined in most cases (e.g. Faure, 1986). Thus, zircons are like elephants: they almost never lose their memory completely.

Any dating of zircons is less valuable if their relative age with respect to fabric elements in the rock is not known; did they come from melt veins or the host rock? If from the host rock, from what type of layering did they come? Careful description, with drawings and photographs, is needed while sampling for zircon analysis (see also remarks on sampling in Chapter 2). Host rock zircons may dissolve in local melts that form in many high-grade rocks, and new zircons may form after solidification. One should therefore be careful not to combine in one analysis a zircon fraction from the host rock and from leucocratic pods which may have been melt pockets.

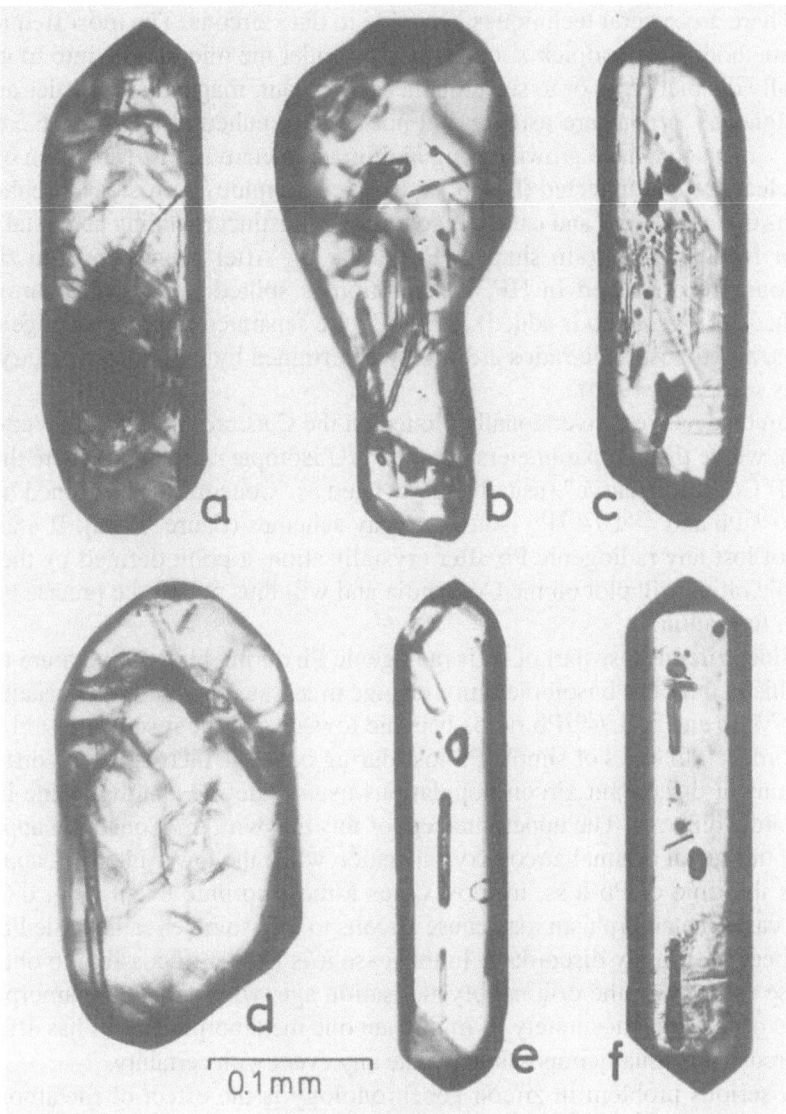

Fig. 6.5. Photomicrographs of several morphologically distinct zircon types. (a) Internally zoned crystal with very dark (U-rich) core, light central part and distinct overgrowth (top portion) around an already corroded grain. (b) Clear and strongly resorbed and corroded grain, typical of high-grade metamorphic terrains, containing tiny apatite needles and fluid inclusions. (c) Near-euhedral grain of igneous origin with well preserved pyramidal surfaces and numerous fluid inclusions (black spots). (d) Clear and rounded detrital grain whose crystallographic surfaces can no longer be identified, enclosing a weakly zoned igneous core derived from a metasedimentary rock. (e) Near-euhedral grain with long-prismatic shape typical of derivation from an igneous source. (f) Euhedral, long-prismatic grain from a metarhyolite with drop-, pipe- and worm-like fluid inclusions.

There are several techniques available to date zircons. The most frequently used method is to handpick zircon fractions under the microscope into morphologically distinct types or to separate them by colour, magnetic properties and/or size. Igneous zircons are usually, but not always, euhedral (e.g. Fig. 6.5 c,e,f) while zircons that have grown during high-grade metamorphism are often ovoid, very clear and multifaceted (Fig. 6.5b, d). Metamorphic "corrosion" of euhedral zircons often occurred and can be recognised by distinct rounding at crystal ends and/or by irregular grain shapes (Fig. 6.5a, b). After separation, the zircon fractions are dissolved in HF, the solution is spiked (a known quantity of enriched radiogenic Pb is added), U and Pb are separated by ion exchange techniques, and the isotopic ratios are finally determined by mass spectrometry (for details see Faure, 1986).

Zircon data are conventionally plotted on the Concordia-Diagram (Wetherill, 1956), where the two parameters are the Pb/U isotopic ratios, and where the so-called "Concordia curve" (usually abbreviated as "Concordia") is defined by the $^{238}U/^{206}Pb$ and $^{235}U/^{207}Pb$ isotopic decay schemes (Faure, 1986). If a zircon has not lost any radiogenic Pb after crystallisation, a point defined by the two isotopic ratios will plot on the Concordia and will thus record the precise age of zircon formation.

Since zircons lose part of their radiogenic Pb during high-temperature metamorphism, their U-Pb isotopic ratios change in a systematic manner (usually the $^{238}U/^{206}Pb$ and $^{235}U/^{207}Pb$ ratios become lower), and the results are said to be "discordant". In cases of simple Pb-loss during only one thermal event, different fractions of discordant zircon populations usually define a straight line in the Concordia diagram. The upper intercept of this line with the Concordia approximates the age of original zircon crystallisation while the lower intercept approximates the time of Pb-loss, in most cases a metamorphic event (Fig. 6.6). In some cases metamorphism may cause zircons to lose so much radiogenic Pb that they become grossly discordant. In this case it is almost impossible to obtain a precise estimate of the original crystallisation age, whereas the metamorphism can be dated rather accurately. If more than one metamorphic event has affected the zircons it is usually impossible to date any event with certainty.

A serious problem in zircon geochronology is the effect of metamorphic overgrowth (e.g. Fig. 6.5a). If overgrowth occurs during a high-grade event the zircon core obviously has an older age than the rim, and analysis of the whole grain, or a mixture of grains, may give geologically meaningless "mixed" ages. This can be overcome, in many cases, by removing the rim through abrasion (e.g. Krogh, 1982). Nevertheless, zircon populations in high-grade metamorphic rocks are almost always heterogeneous, and bulk fraction analysis may not accurately date the time of rock formation.

In view of the complexity of zircon geochronology, modern studies try to overcome the isotopic heterogeneity of large bulk fractions by analysing single grains. For reasonably large grains (200 μm or more) this can now be done with high precision by conventional dissolution (e.g. Lancelot et al., 1976). An alternative technique recently proposed is to evaporate single grains directly in the mass spectrometer without chemical treatment and after wrapping them into Rhenium filaments (Kober, 1986). In this technique only the $^{207}Pb/^{206}Pb$ age can be determined. It is believed that at high temperature only the most stable, origi-nal, radiogenic Pb component is evaporated, so that the $^{207}Pb/^{206}Pb$ ratio measured accurately reflects the time of zircon formation. The method is fairly new, but data so far published compare well with conventional and ion-micro-probe results (e.g. Kober, 1986; 1987; Kober et al., 1989; Kröner et al., 1988).

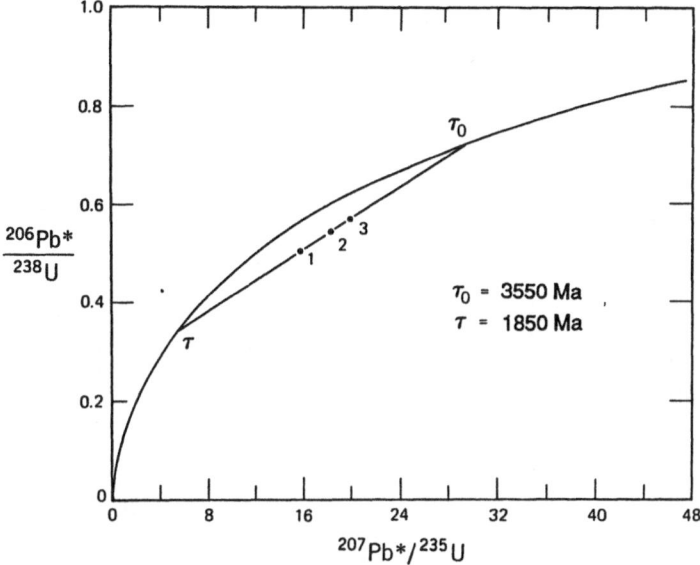

Fig. 6.6. Concordia diagram for three discordant zircon fractions. The discordia line intersects Concordia in two points that date the age of the zircons (t_0) and the age of episodic lead loss (t). From Faure (1986).

The most sophisticated zircon dating technique is by ion microprobe. In this technique a beam of negative oxygen ions sputters positive ions out of a 20-30 µm spot on a polished surface of a zircon grain. These sputtered ions are then analysed in a high-resolution mass spectrometer (Compston et al., 1984). This technique permits the isotopic analysis of minute portions of zircon and has already shown that many zircons, whether igneous or metamorphic, are very heterogeneous with respect to their U-Th-Pb isotopic composition. Metamorphic overgrowth can be readily distinguished from original igneous cores (e.g. Compston & Kröner, 1988), and it is also possible to recognize xenocrystic zircons and those with complex Pb-loss histories. An excellent example of the ion-microprobe technique for the analysis of high-grade metamorphic rocks is provided by Black et al. (1986).

Zircon dating has now reached a precision that makes it possible to constrain the age of fabric formation in a high-grade rock. For example, the layered gneiss shown in Fig. 6.7 is a trondhjemitic orthogneiss whose igneous zircons crys-tallised 984±8 Ma ago. The leucocratic mobilisate clearly cuts the layering and contains igneous zircons with an age of 918±5 Ma. Thus, the structural event giving rise to the pervasive layering must have occurred some time between 918±5 and 984±8 Ma ago. The application of methods like this has shown that many high-grade terrains did not deform synchronously over large regions, but that discrete structural events occurred at different times in different places and produced similar-looking fabrics. One should therefore be cautious in correlating structures over large areas simply by their geometry.

Fig. 6.7. Trondhjemitic gneiss from Hirassagala, Sri Lanka, showing layering cut by leuco‧granitic mobilisate. Age data are derived from single zircon evaporation (Kröner, unpubl.).

It is now also possible to date metamorphic minerals such as garnet, rutile, sphene and monazite at high precision by the U-Pb method (e.g. Schärer et al., 1986; Mezger et al., 1988; 1989) so that metamorphic events and cooling rates can be estimated. This, in combination with igneous crystallisation ages and structural analysis will make it possible in the future to reconstruct virtually the entire history of a high-grade metamorphic terrain from initial transfer into the lower crust to final exhumation and re-exposure at the surface.

7 Origin and Evolution of High-Grade Gneiss Terrains

7.1 Introduction

In the previous chapters we have given an impression of the way in which the analysis of high-grade rocks can be approached in the field. Here we conclude with some current larger-scale ideas on the origin and evolution of such rock assemblages.

High-grade gneiss terrains are considered to represent former portions of the lower continental crust. They make up a significant part of the Precambrian shields, and also occur in many Phanerozoic orogenic belts. They are derived from various proportions of metasedimentary and metavolcanic rocks and granitoid intrusions, and are generally interpreted to be either the result of collision tectonics (analogous to the evolution of the Himalayan orogen, e.g. Dewey & Burke, 1973; Windley, 1981), or the former deep roots of Andean-type continental margins (e.g. Tarney & Windley, 1977). However, such simple models cannot easily explain the complex polymetamorphic and multi-deformational histories of many Precambrian high-grade belts that extend over several hundred million years (e.g. Limpopo belt, southern Africa, Barton & Key, 1981; Napier Complex, Enderby Land, Antarctica, Black et al., 1986; see also listing in Bohlen & Mezger, 1989).

7.2 Two Kinds of Gneiss Terrain

Two major types of high-grade gneiss assemblages have been recognised. One consists predominantly of igneous associations with only minor metasedimentary rocks. The igneous rocks were mainly mafic to felsic volcanic rocks intruded by gabbroic and granitoid gneisses of Tonalite-Trondhjemite-Granodiorite (TTG) composition. They were derived from mafic precursors of probable mantle origin, and are therefore juvenile with no significant crustal prehistory. They are characterised by low $^{87}Sr/^{86}Sr$ initial ratios and positive ε_{Nd} values. The TTG suite has the geochemical and petrographic characteristics of I-type granites. In many of these terrains high-grade metamorphism took place very shortly after crust formation. Part of the late Archaean Lewisian Complex of Scotland is a well studied example of this type and is considered to represent an ancient juvenile (intra-oceanic) island arc complex (Tarney and Windley, 1977; see also papers in Park and Tarney, 1987).

The other type of high-grade gneiss assemblage generally lacks these juvenile components and consists predominantly of clastic and carbonate meta-sedimentary rocks, often clearly of fluviatile or shelf-type origin, intruded by dominantly S-type granitoid rocks. The clastic metasedimentary rocks were mostly derived by erosion of older continental crust, and the granitoid rocks resulted from intracrustal melting, and are thus geochemically and isotopically distinct from the juvenile phases mentioned above (they have high $^{87}Sr/^{86}Sr$ initial ratios and negative ε_{Nd} values). Examples of this type of high-grade terrain are the Limpopo and Namaqua-Natal belts of southern Africa (Tankard et al., 1982), the high-grade rocks of southern India (Drury et al., 1984), the Arunta and Musgrave blocks of central Australia (Shaw et al., 1984), the Inari-Belomoride belt of Finnish Lapland (Barbey et al., 1985), the Grenville belt of eastern North America and its equivalent in Norway (Tobi & Touret, 1985) and the Highland and South West Groups of Sri Lanka (Cooray, 1984). In most of these cases high-grade metamorphism followed long after crust formation, and one of the central unresolved questions is how the supracrustal rocks were transported to the lower crust and then up again. Economically important ore deposits, predominantly Pb-Zn-Ag, occur in this type of high-grade terrain in association with the supracrustal rocks, e.g. Broken Hill and Mt. Isa, Australia, (Marjoribanks et al., 1980) Aggeneys and Black Mountain, Namaqualand, South Africa (Rozendaal, 1986; Ryan et al., 1986) and the Cu-Ni ores of of the Thompson Lake - Moak Lake belt in Canada (Sawkins, 1986).

A useful field guide is that TTG suites with hornblende as the main mafic phase are generally juvenile I-type granitoid rocks, whereas biotite- and/or mus-covite-rich granites are generally of S-type.

7.3 Crustal Structure of Gneiss Terrains

Although many high-grade gneiss terrains have been studied in detail in recent years, we still know very little about the overall structure and composition of the lower continental crust. This is chiefly because much of the continental crust appears to have been extensively sliced up by horizontal tectonic movements, and most exposed gneiss terrains represent relatively thin portions of the crust as a whole. Relatively intact, major sections of the lowermost part of the crust are only exposed in a few places (possibly the Ivrea Zone of the Alps; Sills & Tarney, 1984). Most high-grade terrains appear to represent only partial sections of the middle to upper part of the lower crust (depths of 20-30 km; Bohlen and Mezger, 1989).

One way of obtaining information on the lowermost part of the crust is to study lower crustal xenoliths, brought to the surface in Tertiary to Recent basalt eruptions (e.g. Rudnick and Taylor, 1987). Modern crustal reflection studies have also provided a wealth of new information on the structure of the lower crust (e.g. Oliver, 1978; Smithson et al., 1986; Meissner, 1989). In particular they have shown that near-horizontal reflectors characterise many lower crustal profiles. However, it is still unclear what these reflectors really mean (rock type boundaries, tectonic discontinuities, fluid pathways, or a combination of these features), and until this problem is resolved, seismic profiles provide little detailed information on structural style in the lower crust. Seismic profiles have also revealed many shallow-dipping reflectors that go virtually to the base of the crust, and these are interpreted as major shear zones along which the crust was internally decoupled and lower crustal segments were brought to the surface (e.g. Moine thrust, Scotland, Wind River thrust, Colorado).

7.4 Hypotheses on the Origin and Evolution of Gneiss Terrains

There are currently several competing hypotheses concerning the origin and evolution of high-grade gneiss terrains. They are schematically shown in Fig. 7.1 and are outlined below, following Newton (1987):

(1) *Hot-spot hypothesis*. This model requires a rising hot spot or mantle plume that transports hot mantle material to the base of the crust and results in upper crustal rifting (Fig. 7.1A). Mantle melts are denser than the continental crust and so only a small portion of this melt reaches the surface or intrudes into the crust. The bulk of the material remains at or near the crust-mantle interface and constitutes a so-called "magmatic underplate". Such underplating models have become popular in recent years and may account for a significant proportion of crustal growth, at least in the Precambrian (for discussion and references see Kröner, 1985). The mafic-ultramafic magmas thus produced and stored could provide the heat required for high-grade metamorphism (Bohlen & Mezger, 1989) and the low-P (H_2O) volatiles required to purge the lower crust of water (Fig. 7.1A). In this model, high-grade metamorphism is associated with crustal thinning and extensional deformation rather than compression, and it is difficult to see how sediments once deposited at the surface can be transferred downwards 20-30 km into the middle to lower crust.

(2) *Intracontinental subduction (A-subduction)*. This model is a continuation of model I (Fig. 7.1A) in which mantle plume activity leads to continental extension and basin formation. However, in this case the extension process does not lead to complete continental separation. Instead the cold, subcrustal mantle lithosphere is "delaminated" (i.e. peeled off from the overlying continental crust), and sinks, triggering crustal shortening above and, at the same time, permitting the upward flow of hot mantle asthenosphere (Kröner, 1983). This, together with tectonic thickening through crustal duplex formation (Fig. 7.1B), may trigger high-grade metamorphism and cause the transport of supracrustal rocks to great depth. This model does not require a subduction zone or subduction-related magmatic processes.

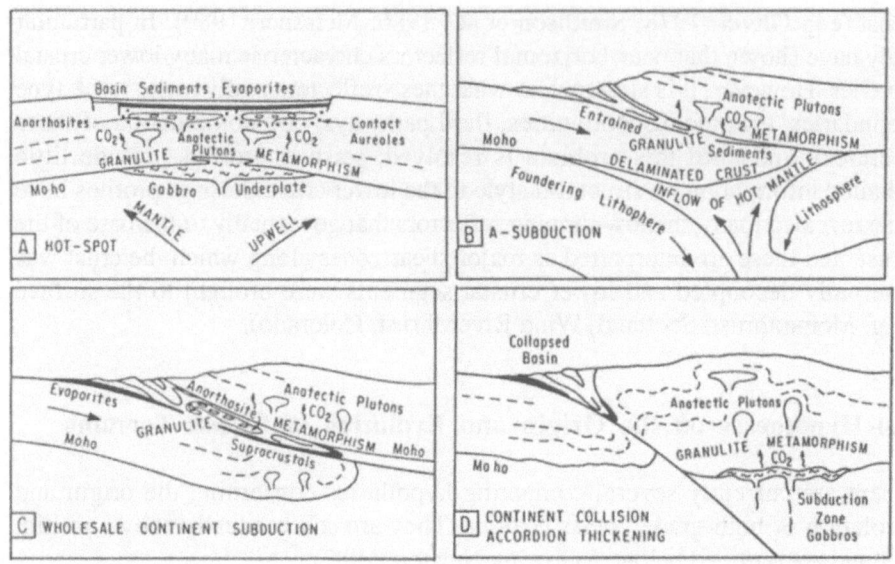

Fig. 7.1. Various tectonic settings producing high-grade metamorphic rocks (after Newton, 1987). See text for explanation.

(3) *Continent-scale underthrusting.* In this model a complete section of continental crust is extensively thrust beneath another segment of continental crust, effectively doubling the continental thickness. A considerable amount of continental shelf-type sedimentary rocks can be entrained in the relatively flat continent-continent interface and transported to great depth, and most of the structures in this compressional regime would also be essentially flat (Fig. 7.1C). This model also provides an uplift mechanism in the isostatic rebound of a doubly-thickened crust (Ellis, 1987).

(4) *Continental collision with accordion-style thickening.* This is the well-known model of Dewey and Burke (1973) applied to the collision of India with Asia and the evolution of the Himalayas. Continental underthrusting and associated thrust and nappe-stacking results in an overthickened crust (Mattauer, 1986), in the lower parts of which high-grade metamorphism occurs in a compressional regime (Fig. 7.1D). Zen (1988) has modelled the thermal history of this scenario and concluded that the characteristic time lapse for high-grade metamorphism and anatexis after tectonic stacking is several tens of millions of years. Supracrustal rocks originally deposited on passive or active margins of the converging continents may be transferred to great depths during the collision process. Magmas rising from a subduction zone below the underriding continent could have provided the heat required to initiate high-grade metamorphism.

Bohlen & Mezger (1989) argue that xenoliths from the real lower crust only rarely consist of supracrustal rocks and that there is a general increase in the mafic character of the xenoliths with depth. This observation seems to support substantial vertical crustal growth by magma underplating, and the above authors suggest that episodic underplating and subsequent differentiation of magma batches account for the long metamorphic histories observed in many high-grade terrains.

Many high-grade terrains provide evidence for considerable horizontal shortening predating the thermal peak of regional metamorphism, and this deformation is often accompanied by intrusion of voluminous granitoid sheets. This "intraplating" effect can also lead to substantial crustal thickening and may be accompanied by tectonic stacking (Myers, 1976). The most likely tectonic setting for such a scenario is an active continental margin (Kay & Kay, 1981), and this should produce long, linear belts of high-grade metamorphic rocks as indeed found in many late Proterozoic and Phanerozoic orogens. The extensive and sediment-dominated high-grade gneiss terrains in the ancient shields such as in southern India and East Africa, however, are difficult to explain by the continental margin underplating model. Thompson (1989) has advanced the attractive hypothesis that in such areas the crust is first thinned during several tens of millions of years by extension, leading to basin formation and deposition of clastic sediments, followed by a period of intense homogeneous shortening that may lead to doubling or even tripling of the previously thinned crust by tectonic stacking. Thompson (1989) also noted that tectonic stacking would tend to thicken the crust mainly downward with respect to sea level rather than upward, thus making it possible for supracrustal sequences to be transported to considerable crustal depth. The wedge-shaped form of seismic reflectors recently discovered in several lower crustal profiles and named 'crocodiles' by Meissner (1989) may be an expression of this stacking process.

A good example for the formation of such high-grade terrains may be the ongoing collision of India and Asia. The ancient northern Indian continental margin, previously thinned during the Gondwana breakup process, consisted of Archaean to Palaeozoic igneous and supracrustal rocks. After the India-Asia suturing 40-50 Ma ago, this margin continued to be thrust under the Asian continent during the last 40 Ma, and about 1500 km of the Indian continental lithosphere has now disappeared below Tibet (Mattauer, 1986) and may be presently transformed into an extensive high-grade terrain.

It is likely that there is more than just one mechanism that produces high-grade rocks in the continental crust, and a hypothesis now popular may be discarded in the future on the basis of better or more complete data. Detailed struc-tural work aiming at reconstructing the deformation history of high-grade terrains, combined with petrology, geochemistry, isotopes, precise geochronology and high-resolution seismology, are likely to lead to new insights and require close collaboration between the various fields in the earth sciences. Understanding the evolution of the lower and middle crust remains a major scientific challenge for the last decade of this century.

8 Problem Section

8.1 Introduction

This manual describes some of the most important features that should be observed, recorded and analysed in the field interpretation of gneiss terrains. However, every outcrop is unique, and unforeseen problems will be encountered. Increasing experience will make it easier to solve these problems but will also enable the detection of further subtle complexities. In this section we present examples of some commonly encountered phenomena that may easily be misinterpreted. These are followed by a series of exercises in the form of outcrop maps and photographs.

8.2 Informative Misinterpretations

In previous sections we have outlined some of the better known problems that are often encountered in gneiss terrains. Such an exercise can never be complete, and our main purpose here is to sharpen the reader's eye for potential problems and ways to solve them. Gneisses are difficult rocks to work with, and much of their history is not clearly apparent. We outline below a number of phenomena that have been misinterpreted.

Two veins making an angle of 2^0 were interpreted as identical and parallel, whereas they were originally orthogonal and cross-cutting before deformation (Figs. 3.20; 4.33). This potential error might have been avoided if the slight obliquity had been noted and if other outcrops had been mapped where, possibly because of lower strain, cross-cutting veins or original obliquity of veins could still have been observed.

A sequence of granitoid rocks containing abundant quartzo-feldspathic veins was isoclinally folded and cut by layer-parallel shear zones. The resulting layered gneisses were interpreted as a stratified sedimentary sequence and a detailed stratigraphy and sedimentary facies model was published.

A rapid change in deformation state within a single rock unit was not exposed or was not observed. The strongly deformed version of the rock was interpreted as older than the less deformed portion of the same rock (Figs. 3.10; 3.11).

A series of intrusive granitoid sheets were interpreted as belonging to two separate intrusion events with older veins being deformed before the emplacement of younger, post-tectonic veins. In fact, however, all the veins were intruded before the deformation but thinner and finer grained intrusive sheets were more intensely deformed, and therefore became more strongly foliated, than older, thicker, coarser grained sheets.

Samples from three parallel layers of metamorphosed igneous rocks - black, grey and white - were taken for whole rock chemical analysis. The resulting data were interpreted as showing a fairly continuous differentiation sequence from black to grey to white, and the rocks were considered to be comagmatic. An alternative interpretation was presented that the grey rock formed by metasomatic alteration of the black rock. However, regional mapping traced the layers into an area of lower strain and revealed that the layers were originally *not* parallel, and that the white rock veined the black one, and that both were cut by dykes of the grey rock.

Samples from a metamorphic gneiss dome were taken at regular intervals to determine the range of P-T conditions over the dome, and to find the geothermal gradient at peak metamorphic conditions. However, the mineral assemblages that define the isograds in the area are not all of the same age; some predate the main phase of deformation, and some postdate this phase. The calculated uplift path and geothermal gradient at peak metamorphic conditions were significantly off the real values.

A suite of aplitic dykes was sampled for Rb-Sr age determination. Field evidence of the presence of one isoclinally folded and one undeformed set of aplites was missed. When the isotopic results were plotted some points fell on an isochron and some were way off. Some points were derived from the older dykes and some belonged to the second set of veins, but without knowledge of the field relations the scatter of the data could not be interpreted.

A gneiss was metamorphosed in granulite facies and developed abundant small melt patches in which many new euhedral zircons grew. The gneiss was subsequently retrograded and deformed in amphibolite facies, and the melt patches were streaked out into the main new tectonic layering. Zircons taken from the gneiss (for either single or multiple grain U-Pb geochronology) and thought to give the igneous age of the rock were dominantly euhedral zircons from the melt patches and gave the age of granulite facies metamorphism, rather than the igneous age of the gneiss protolith.

8.3 Problems

The following map (Fig. 8.1) and photographs (Figs. 8.2-8.16) show a number of outcrops of high-grade gneiss terrains from different regions. The reader should try to establish a sequence of events in each case without consulting the text. We have given our solution at the end of this chapter.

126

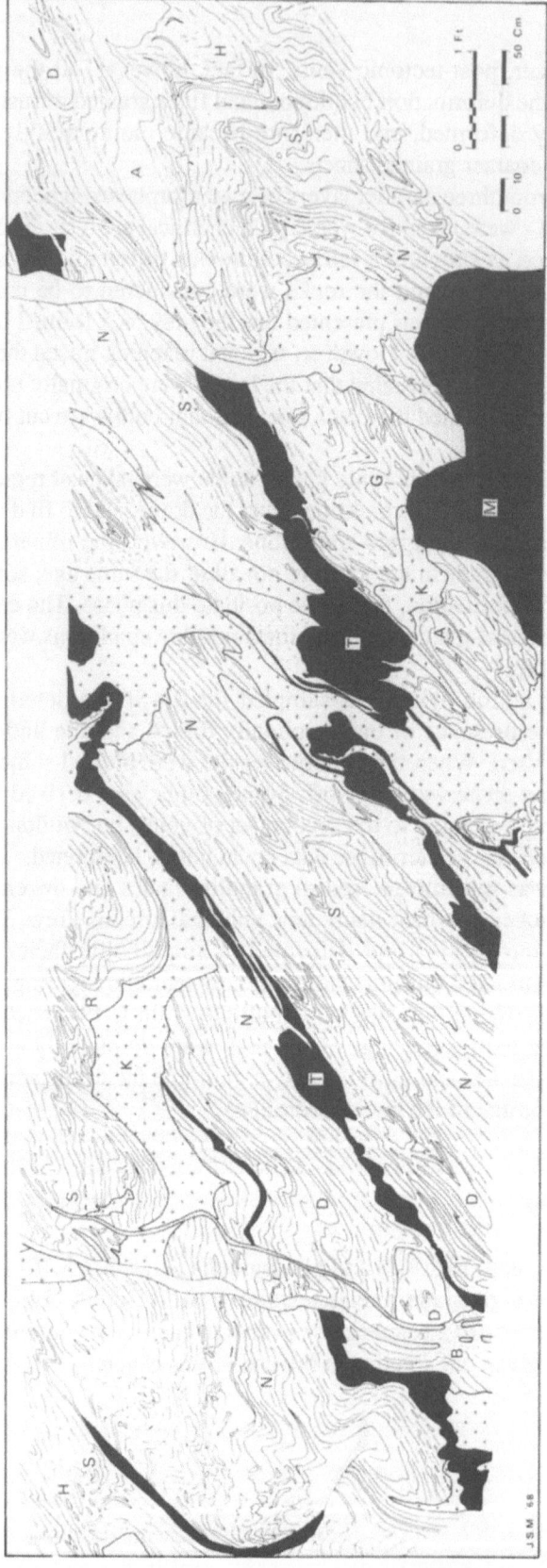

Fig. 8.1. Smooth outcrop surface showing a complex geological history of intrusive veins and deformation, Scarp, Harris, Outer Hebrides, Scotland. Work out the sequence of events. It is easiest to work backwards, starting with the youngest features. When the sequence is solved, it is then usually described as it evolved from the oldest to the youngest. The different rock types are indicated by the following ornamentation: pegmatite of various ages - black; amphibolite - stippled; granite - large crosses; aplitic granite - small crosses; quartzo-feldspathic veins - white; quartz-feldspar-biotite gneiss is shown by discontinuous lines, which represent foliation and/or layering.

Fig. 8.2. Ameralik, SW Greenland.

Fig. 8.3. West Greenland.

128

Fig. 8.4. Fiskenæsset, SW Greenland.

Fig. 8.5. Limpopo belt, South Africa.

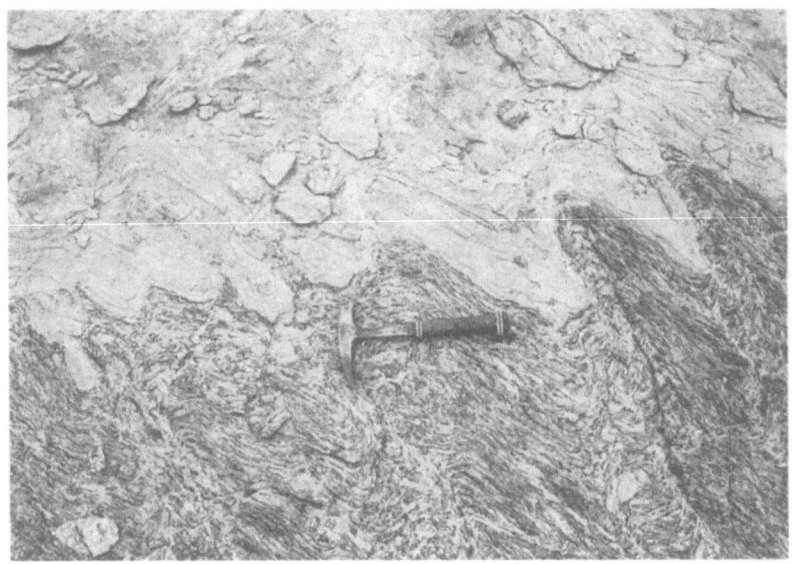

Fig. 8.6. Central Damara belt, Namibia.

Fig. 8.7. Fiskenæsset, SW Greenland.

Fig. 8.8. Qin Ling Mountains, central China.

Fig. 8.9. Northern Somalia.

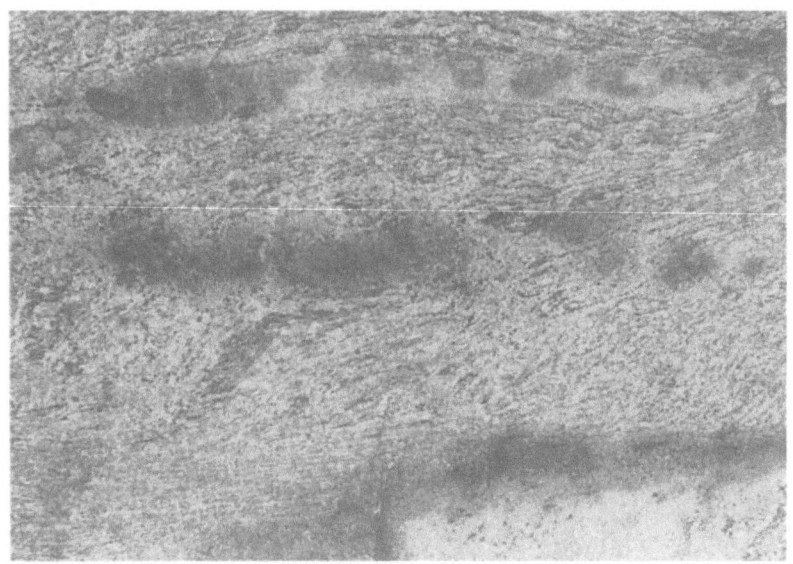

Fig. 8.10. Kurunegala, central Sri Lanka.

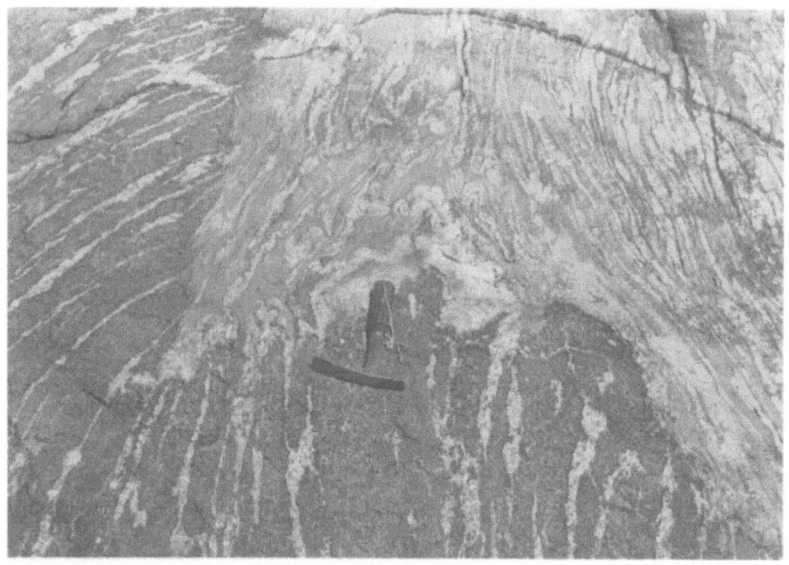

Fig. 8.11. Leeuwin Complex, Proterozoic Pinjarra Orogen, SW Australia.

Fig. 8.12. Southern Finland.

Fig. 8.13. Leeuwin Complex, SW Australia.

Fig. 8.14. Fiskenæsset, SW Greenland.

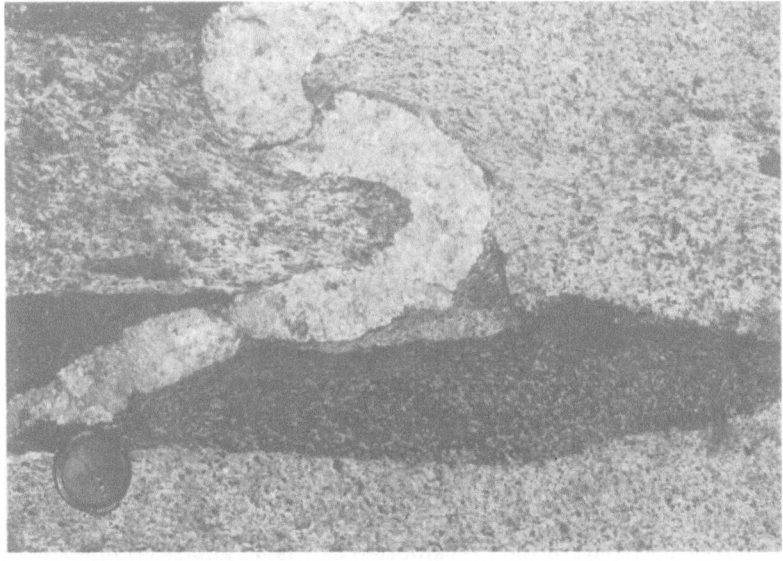

Fig. 8.15. Crystallina, Ticino, Switzerland.

Fig. 8.16. Ameralik, SW Greenland.

Interpretation

Fig. 8.1. The oldest rocks are layered gneiss (N) containing layers and lenses of quartzo-feldspathic pegmatite (S) and layers, lenses and dismembered fold cores of amphibolite (D). The latter could be either the very oldest rocks or disrupted early mafic dykes. These layers have been tightly folded by deformation d1, and in most cases are now seen as isolated rootless folds with boudinaged limbs.

These rocks and structures are cut by quartzo-feldspathic veins (A) and aplitic granite veins (R). They are in turn cut by granite veins (K) and pegmatite veins (T). (K) and (T) are mutually crosscutting and apparently of the same age. During a subsequent episode of deformation, d2, the pegmatite veins (T) were ptygmatically folded and small tight folds of a second phase (H) developed in the gneiss. The latter formed fold interference structures with d1 folds (G). In some places (L) d2 folds possess an axial planar foliation defined by biotite. All these rocks and structures are cut by amphibolite dykes (C), which are associated with displacement of older veins and layering; there must either be a shear component on the dykes, or they intruded along faults or shear zones. The dykes (C) were then boudinaged in association with partial melting (B) of the older gneiss. The youngest rock is pegmatite (M) that cuts all the other rocks and structures.

These rocks are part of the Precambrian Lewisian gneiss complex of Scotland. The outcrop occurs on the coast of the island of Scarp, off the Isle of Harris in the Outer Hebrides (Myers, 1970). It is part of a small area of relatively low finite strain. Throughout most of the region the rocks are much more deformed, the different components are streaked out into parallelism, and most of the evidence of the early geological history shown here is obliterated (cf. Fig. 4.1b). Thus, although the outcrop is tiny, it is very important, and such outcrops should be identified and described in detail.

Fig. 8.2. Strongly deformed heterogeneous, layered quartzo-feldspathic gneiss with dark and light components and isoclinally folded and boudinaged, partly fragmented amphibolite dykes all brought into parallelism, then cut by aplite veins. 3.65 Ga Amitsoq gneiss and Ameralik dykes. Ameralik, SW Greenland.

Fig. 8.3. The centre of the photograph shows a layered gneiss with amphibolite units, some of which are oblique to others, indicating that at least some of them are intrusive in origin. The amphibolites and the layered gneiss in the rest of the outcrop were strongly flattened (indicated by the internal shape fabric) and the amphibolite layers were boudinaged. Some boudinage is also visible in the gneiss at lower left. Because quartzo-feldspathic material fills the boudin necks, the boudinage was probably associated with local partial melting. Subsequently, the layering was folded and a ductile shear zone developed (top), truncating the layering. The age relation of the folding and the shear zone is unclear. West Greenland.

Fig. 8.4. Undeformed leucogabbro with relic cumulus plagioclase megacrysts (at left, below hammer) is in sharp contact with strongly deformed tectonically layered leucogabbro, at right. Possible interpretations are: (1) the massive leucogabbro was intruded into the layered leucogabbro subparallel to the foliation; (2) the layered leucogabbro is derived from the massive leucogabbro by strong ductile deformation. Close observation shows that some of the feldspar megacrysts in the contact zone grade into lenses in the layered leucogabbro and there is a rapid transition from massive to layered leucogabbro, and no sharp truncation of the layered rock by the massive rock. In fact, the layered leucogabbro is in a shear zone; deflection (as in Fig. 4.13) would be seen for a movement direction at a small angle to the outcrop surface, but not necessarily if the movement direction was steep. The latter is probably the case here. Fiskenæsset, SW Greenland.

Fig. 8.5. A layered gneiss (bottom) contains some aplitic veins at an angle to layering. The layering and a mafic dyke (at lower right, parallel to the layering - the age relation is uncertain) are truncated by a younger dolerite dyke (below the hammer), with an aplitic vein in the centre. Although the later dyke appears to be undeformed, boudinage of the late aplite indicates that it was subject to considerable stretching. The older mafic dyke is dated at 3.6 Ga; the younger dyke at 3.0 Ga. Sand River Gneiss locality, Limpopo belt, South Africa.

Fig. 8.6. A folded contact between a strongly foliated augen gneiss and a metaquartzite. The augen gneiss is a 2 Ga old igneous rock; the quartzite is ca. 800 Ma old. This unimpressive looking contact is therefore an original unconformity or a major fault plane. Central Damara belt, Namibia.

Fig. 8.7. Folded pegmatite-layered gneiss, with migmatitic veins parallel to the axial planes. Two interpretations are possible: (1) the veins are of the same age or younger than the folding, or (2) they were present at a high angle to the shortening direction which formed the folds. In the latter case, the veins would be expected to show boudinage or folding in at least some outcrops. Fiskenæsset, SW Greenland.

Fig. 8.8. Dome-and-basin structures? Sheath folding? Neither - simply a topographic effect of an irregular outcrop surface with tightly folded aplite veins. Width shown in photograph is 50 cm. Qinling Mountains, central China.

Fig. 8.9. An aplite dyke cuts across a foliation in a homogeneous gneiss, but shows a curious jog. There are two possible interpretations: (1) a narrow shear zone which parallels the foliation in the gneiss cuts the dyke in the centre; (2) the dyke formed with its present geometry, and opened parallel to the jog in the centre, which acted as a transform fault' as in a midoceanic ridge. The latter seems to be the case here because no trace of a late shear zone is visible in the surrounding gneiss. This can be checked, for the aplite in the jog should be undeformed; a sample should be collected to check for traces of ductile deformation. Northern Somalia.

Fig. 8.10. Pale gneiss with medium-grade mineral assemblage and H_2O-rich fluid inclusions, surrounding dark patches with high-grade (hypersthene-bearing) assemblage and fluid inclusions containing CO_2. The dark patches seem to follow a shear zone at the bottom of the photograph but also occur as a band in the gneiss in the middle of the photograph parallel to the shear zone, and in a quartzo-feldspathic layer at the top. Possible interpretations are: (1) CO_2 fluxing through the rock changed mineral assemblages locally. The occurrence of dark patches in the shear zone seems to support this, but the occurrence of dark patches elsewhere should also be explained; (2) retrograde metamorphism caused the breakdown of high-grade assemblages in most of the rock except in some elongate patches, perhaps due to local variations in porosity, composition or grain size. The aplite vein and the shear zone could both be examples of such variations. Only detailed analysis of a large number of outcrops in the area may solve the problem. Width shown in photograph is 1 m. Kurunegala, central Sri Lanka.

Fig. 8.11. The complex fabric in this outcrop must be due to inhomogeneous deformation. The contact between black amphibolite (derived from gabbro) and grey gneiss (derived from granite) is folded, and the style of folding indicates a competency contrast between the two rock units. Pegmatite veins occur in both rock units, and their orientation is a function of their position in the fold. The veins were refracted across the boundary between the amphibolite and the gneiss, and were then deformed. The pegmatite veins in the amphibolite and some of those in the gneiss appear to have been intruded along a foliation associated with the fold, but some veins in the gneiss appear to be older than this folding. The sequence derived from this outcrop would therefore be: intrusion of gabbro and granite with pegmatite veins (sequence not clear); intrusion of pegmatite veins during deformation that formed the fold. The fold is parasitic on a large-scale isoclinal fold (not visible in this outcrop). Metamorphism reached high-grade conditions after this folding and was followed by retrogression to medium-grade conditions. Leeuwin Complex, Proterozoic Pinjarra Orogen, SW Australia.

Fig. 8.12. Shortened and folded boudins of an amphibolite sill in metavolcanic rocks. Deformation of amphibolite xenoliths could cause this kind of structure but the small pegmatite patch enclosed by the 'pincer-shaped' amphibolite block, and the thin 'tail' connecting two blocks are good evidence for an early episode of boudinage. Haraholm, Åland, southern Finland.

Fig. 8.13. Black amphibolite derived from gabbro, grey amphibolite derived from leucogabbro, and a quartz vein (white in upper part of photograph) whose early relations are no longer recognisable. They have been intensely deformed together and developed a foliation parallel to compositional layering. Blotchy feldspathic pegmatite veins (white in lower part of photograph) cut across the tectonic layering and developed after the peak of deformation during high-grade recrystallisation. Width shown in photograph is 1 m. Leeuwin Complex, SW Australia.

Fig. 8.14. This type of outcrop in a gneiss terrain is very illustrative since a large part of the history of the rock can be read from a few square metres. The sequence of events is as follows: (1) black amphibolite derived from basalt was deformed and veined by pegmatite (some early veins are parallel to a foliation, whereas some crosscut the early veins and foliation); (2) amphibolite was fragmented by the intrusion of pale tonalite (now quartzo-feldspathic gneiss); (3) the rocks were strongly deformed: some amphibolite was streaked out into layers (e.g. below the hammer) and some larger blocks of amphibolite were rotated but little deformed internally; (4) homogeneous grey dykelets of tonalite intruded during moderate ductile deformation and all the rocks were recrystallised. Fiskenæsset, SW Greenland.

Fig. 8.15. An amphibolite xenolith in a granodiorite is cut by a pegmatite vein, and both have been deformed. There is a weak shape fabric in the granodiorite, slightly fanning in the axial plane of the folded pegmatite. This indicates that the foliation is associated with the growth of the fold. The pegmatite is boudinaged in the fold limb where it cuts the xenolith. Because the long axis of the deformed xenoliths, the shape fabric and the axial plane of the fold coincide, it is most likely that the pegmatite cut the granodiorite and the xenoliths before any deformation. It is possible, however, that the xenoliths suffered some deformation before the intrusion of the pegmatite. Strain values in the xenolith and in the pegmatite must be compared to solve this problem. Crystallina, Ticino, Switzerland.

Fig. 8.16. Strongly foliated gneiss with flattened amphibolite blocks cut by an undeformed pegmatite vein. The homogeneous fabric of the host gneiss indicates that the gneiss is probably of igneous origin. The amphibolite fragments could have been derived from a mafic dyke that was boudinaged and then flattened, or formed by flattening of amphibolite xenoliths. Regional mapping has shown that the former is the case here. Determination of the absolute age of the host gneiss and the pegmatite vein would give age limits for the deformation. Ameralik, SW Greenland.

9 References

Allen AR (1979) Mechanism of frictional fusion in fault zones. J Struct Geol 1: 231-243

Anovitz LM, Essene EJ (1987) Compatibility of geobarometers in the system CaO-FeO-Al_2O_3-SiO_2-TiO_2 (CFAST): implications for garnet mixing models. J Geol 95: 633-645

Aranovich LYa, Podlesskii KK (1983) The cordierite-garnet-sillimanite-quartz equilibrium experiments and applications. In: Saxena SK (ed) Kinetics and equilibrium in mineral reactions. Springer, Berlin Heidelberg New York, pp 173-198

Arndt NT, Goldstein SL (1987) Use and abuse of crust-formation ages. Geology 15: 893-895

Ashworth JR (ed) (1985) Migmatites. Blackie, Glasgow

Barbey P, Convert J, Moreau B, Capdevila R, Hameurt J (1985) Petrogenesis and evolution of an early Proterozoic collisional orogenic belt, the granulite belt of Lapland and the Belomorides (Fennoscandia). Geol Soc Finland Bull 56: 161-188

Barnes JW (1981) Basic geological mapping. Open University Press, Milton Keynes, 112 pp

Barnicoat A (1983) Metamorphism of the Scourian complex, NW Scotland. J. Metamorphic Geol. 1: 163-182

Barton Jr JM, Key R (1981) The tectonic development of the Limpopo mobile belt and the evolution of the Archaean cratons of southern Africa. In: Kröner A (ed) Precambrian plate tectonics. Elsevier, Amsterdam, pp 185-212

Behrmann JH, Mainprice D (1987) Deformation mechanisms in a high-temperature quartz-feldspar mylonite: evidence for superplastic flow in the lower continental crust. Tectonics 140: 297-305

Bell TH, Etheridge MA (1973) Microstructure of mylonites and their descriptive terminology. Lithos 6: 337-348

Berthé D, Choukroune P, Jegouzo P (1979) Orthogneiss, mylonite and non-coaxial deformation of granites: the example of the south-Armorican shear zone. J Struct Geol 1: 31-42

Best MG (1982) Igneous and metamorphic petrology. Freeman, San Francisco, 630 pp

Black LP, McCulloch MT (1987) Evidence for isotopic equilibration of Sm-Nd whole-rock systems in early Archaean crust of Enderby Land Antarctica. Earth Planet Sci Lett 82: 15-24

Black LP (1988) Isotopic resetting of U-Pb zircon and Rb-Sr and Sm-Nd whole-rock systems in Enderby Land Antarctica: implications for the interpretation of isotopic data from polymetamorphic and multiply deformed terrains. Precambrian Res 38: 355-366

Black LP, Williams IS, Compston W (1986) Four zircon ages from one rock - the history of a 3930-Ma-old granulite from Mount Sones, Enderby Land Antarctica. Contrib Mineral Petrol 94: 427-437

Blumenfeld P (1983) Le tuilage des megacristeaux; un critiere d'ecoulement rotationnel. Bull Soc Geol France 25: 309-318

Bohlen SR (1987) Pressure-temperature-time paths and a tectonic model for the evolution of granulites. J Geol 24: 617-632

Bohlen SR, Wall VJ, Boettcher AL (1983) Geobarometry in granulites. In: Saxena SK (ed) Kinetics and equilibrium in mineral reactions. Springer, Berlin Heidelberg New York, pp 41-172

Bohlen SR, Mezger K (1989) Origin of granulite terranes and the formation of the lowermost continental crust. Science 244: 326-329

Bouchez JL, Lister GS, Nicolas A (1983) Fabric asymmetry and shear sense in movement zones. Geol Rundschau 72: 401-419

Bridgwater D, Keto L, McGregor VR, Myers JS (1976) Archaean gneiss complex of Greenland. In: Escher A, Watt WS (eds) Geology of Greenland. Grønlands Geol Unders, Copenhagen, pp 18-75

Burri C (1959) Petrochemische Berechnungsmethoden auf äquivalenter Grundlage. Birkenhäuser, Basel, 334 pp

Chatterjee ND, Johannes W (1974) Thermal stability and standard thermodynamic properties of synthetic 2N1-muscovite $KAl_2(AlSi_3O10)(OH)_2$. Contrib Mineral Petrol 48: 89-114

Clemens JD, Vielzeuf D (1987) Constraints on melting and magma production in the crust. Earth Planet Sci Lett 86: 287-306

Cobbold PR, Quinquis H (1979) Development of sheath folds in shear regimes. J Struct Geol 2: 119-126

Compston W, Williams IS, Meyer C (1984) U-Pb geochronology of zircons from Lunar Breccia 73217 using a sensitive high mass-resolution ion microprobe. J Geophys Res 89 (Suppl): B525-B534

Compston W, Kröner A (1988) Multiple zircon growth within early Archaean tonalitic gneiss from the Ancient Gneiss Complex, Swaziland. Earth Planet Sci Lett 87: 13-28

Cooray PG (1984) An introduction to the geology of Sri Lanka (Ceylon), 2nd edn. National Museums of Sri Lanka Publication, 340 pp

Currie KL, Ferguson J (1970) The mechanism of intrusion of lamprophyre dikes indicated by 'offsetting' of dikes. Tectonophysics 9: 525-535

DePaolo DJ (1981) Neodymium isotopes in the Colorado Front Range and crust-mantle evolution in the Proterozoic. Nature 291: 193-196

DePaolo DJ (1988) Neodymium isotope geochemistry. Springer, Berlin Heidelberg New York, 187 pp

DePaolo DJ, Manton WI, Grew ES, Halpern M (1982) Sm-Nd, Rb-Sr and U-Th-Pb systematics of granulite facies rocks from Fyfe Hills, Enderby Land Antarctica. Nature 298: 614-619

Dewey JF, Burke KCA (1973) Tibetan Variscan and Precambrian basement reactivation: products of continental collision. J Geol 81: 683-692

Dodson MH (1979) Theory of cooling ages In Jäger, E, Hunziker, JC (eds) Lectures in isotope geology. Springer, Berlin Heidelberg New York, pp 194-202

Drury SA, Harris NBW, Holt RW, Reeves-Smith GJ, Wightman RT (1984) Precambrian tectonics and crustal evolution in southern India. J Geol 92: 3-20

Eisbacher GH (1970) Deformation mechanics of mylonite rocks and fractured granites in Cobequid Mountains, Nova Scotia, Canada. Bull Geol Soc Am 81: 2009-2020

Ellis DJ (1987) Origin and evolution of granulites in normal and thickened crust. Geology 15: 167-170

Ellis DJ, Green DH (1979) An experimental study of the effect of Ca upon garnet-clinopyroxene Fe-Mg exchange equilibria. Contrib Mineral Petrol 71: 13-22

Ellis DJ, Green DH (1985) Garnet-forming reactions in mafic granulites from Enderby Land Antarctica: implications for geothermometry and geobarometry. J Petrol 26: 633-662

Escher A, Escher JC, Watterson J (1975) The reorientation of the Kangamiut dyke swarm, West Greenland. Can J Earth Sci 12: 158-173

Escher A, Jack S, Watterson J (1976) Tectonics of the North-Atlantic Proterozoic dyke swarm. Phil Trans R Soc London A 280: 529-539

Etheridge MA (1983) Differential stress magnitudes during regional deformation and metamorphism. Geology 11: 231-234

Faure G (1986) Principles of isotope geology, 2nd edn. John Wiley & Sons, New York, 589 pp

Ferry JM, Spear FS (1978) Experimental calibration of the partitioning of Fe and Mg between biotite and garnet. Contrib Mineral Petrol 66: 113-117

Fiorentini E, Hoernes S, Hoffbauer R (1990) Nature and scale of fluid-rock exchange in granulite grade rocks of Sri Lanka: a stable isotope study. In: Vielzeuf, D Vidal, P (eds) Granulites and crustal differentiation. Reidel, Dordrecht, NATO ASI Series C: Mathematial and Physical Sciences

Francis PW, Sibson RH (1973) The Outer Hebrides thrust. In: Park RG, Tarney J. (eds).The early Precambrian of Scotland and related rocks of Greenland. Conference Proceedings, Univ. Keele, Staffordshire, UK, pp 95-104

Fry N (1984) The field description of metamorphic rocks. Geol Soc Handbook Series, London

Gapais D, White SH (1982) Ductile shear bands in a naturally deformed quartzite. Text Microstr 5: 1-17

Graham CM, Powell RA (1984) Garnet-hornblende geothermometer: calibration testing and application to the Pelona schist, southern California. J Metam Geol 184: 13-31

Gray CM, Oversby VM (1972) The behaviour of lead isotopes during granulite facies metamorphism. Geochim Cosmochim Acta 36: 939-952

Grew ED (1984) A review of Antarctic granulite-facies rocks. Tectonophysics 105: 177-191

Grocott J (1977) The relationship between Precambrian shear belts and modern fault systems. J Geol Soc London 133: 257-262

Grocott J (1981) Fracture geometry of pseudotachylyte generation zones: a study of shear fractures formed during seismic events J Struct Geol 3: 169-179

Halls AC, Fahrig WF (eds.) (1987) Mafic dyke swarms. Geol Assoc Canada Spec Publ 34: 503 pp

Hansen EC, Janardhan AS, Newton RC, Prame WKBN, Ravindra Kumar AG (1987) Arrested charnockite formation in southern India and Sri Lanka. Contrib Mineral Petrol 96: 225-244

Hansen EC, Newton RC,Janardhan AS (1984) Fluid inclusions in rocks from amphibolite facies gneiss to charnockite progression in southern Karnataka, India: direct evidence concerning the fluids of granulite metamorphism. J Metam Geol 2: 249-264

Harley SL, Green DH (1982) Garnet-orthopyroxene barometry for granulites and peridotites. Nature 300: 697-701

Harley SL (1984a) The solubility of alumina in orthopyroxene coexisting with garnet in FeO-MgO-Al$_2$O$_3$-SiO$_2$ and CaO-FeO-MgO-Al$_2$O$_3$-SiO$_2$. J Petrol 25: 665-696

Harley SL (1984b) An experimental study of the partitioning of Fe and Mg between garnet and orthopyroxene. Contrib Mineral Petrol 86: 359-373

Harley SL (1987) Precambrian geological relationships in high-grade gneisses of the Rauer Islands East Antarctica. Aust J Earth Sci 34: 175-207

Harley SL (1988) Proterozoic granulites from the Rauer Group East Antarctica I. Decompression pressure-temperature paths deduced from mafic and felsic gneisses. J Petrol 29: 1059-1095

Harley SL (1989) The origins of granulites: a metamorphic perspective. Geol Mag 126: 215-247

Harris NBW, Holland TJB (1984) The significance of cordierite-hypersthene assemblages from the Beitbridge region of the central Limpopo belt: evidence for rapid decompression in the Archaean? Am Mineral 69: 1036-1049

Hensen BJ, Green DH (1973) Experimental study of the stability of cordierite and garnet in pelitic compositions at high pressures and temperatures. III Synthesis of experimental data and geological applications. Contrib Mineral Petrol 38: 151-166

Hensen BJ (1987) P-T Grids for silica-undersaturated granulites in the system MAS(n+4) and FMAS(n+3)-tools for the derivation of P-T paths of metamorphism. Contrib Mineral Petrol 5: 255-271

Hobbs BE, Means WD, Williams PF (1976) An outline of structural geology. John Wiley & Sons, New York, 570 pp

Hobbs BE, Ord A, Teyssier C (1986) Earthquakes in the ductile regime? Pure Appl Geophys 124: 309-336

Hobbs BE, Ord A (1988) Plastic instabilities: implications for the origin of intermediate and deep focus earthquakes. J Geophys Res 93: 10521-10540

Holdaway MJ, Lee SM (1977) Fe-Mg cordierite stability in high-grade pelitic rocks based on experimental, theoretical and natural observations. Contrib Mineral Petrol 63: 175-198

Holdaway MJ (1971) Stability of andalusite and the aluminum silicate phase diagram. Am J Sci 272: 91-131

Hoppe G (1963) Die Verwendbarkeit morphologischer Erscheinungen an akzessorischen Zirkonen für petrologische Auswertungen. Abh Dtsch Akad Wiss Berlin pp 1-130

Huang WL, Wyllie PJ (1975) Melting reactions in the system NaAlSi3O8-KAlSi3O8-SiO2 to 35 kilobars, dry and with water. J Geol 83: 737-748

Hudleston PJ (1989) The association of folds and veins in shear zones. J Struct Geol 11: 949-957

Kay RW, Kay SM (1981) The nature of the lower continental crust: inferences from geophysics, surface geology and crustal xenoliths. Rev Geophys Space Phys 19: 271-297

Kerrich R, Allison I, Barnett RL, Moss S, Starkey J (1980) Microstructural and chemical transformations accompanying deformation of granite in a shear zone at Mieville Switzerland with implications of stress corrosion cracking and superplastic flow. Contrib Mineral Petrol 73: 221-242

Kober B (1986) Whole-grain evaporation for $^{207}Pb/^{206}Pb$-age investigations on single zircons using a double filament thermal ion source. Contrib Mineral Petrol 93: 482-490

Kober B (1987) Single-zircon evaporation combined with Pb+ emitter-bedding for $^{207}Pb/^{206}Pb$-age investigations using thermal ion mass spectrometry, and implications to zirconology. Contrib Mineral Petrol 96: 63-71

Kober B, Pidgeon RT, Lippolt HJ (1989) Single-zircon dating by stepwise Pb-evaporation constrains the Archaean history of detrital zircons from the Jack Hills, Western Australia. Earth Planet Sci Lett 91: 286-296

Koziol AM, Newton RC (1988) Redetermination of the anorthite breakdown reaction and improvement of the plagioclase-garnet-Al2SiO5-quartz geobarometer. Am Mineral 73: 216-223

Krogh TE (1982) Improved accuracy of U-Pb zircon ages by the creation of more concordant systems using an air abrasion technique. Geochim Cosmochim Acta 46: 637-649

Kröner A (1983) Proterozoic mobile belts compatible with the plate tectonic concept. Geol Soc Am Mem 161: 59-74

Kröner A (1985) Evolution of the Archean continental crust. Ann Rev Earth Planet Sci 13: 49-74

Kröner A, Compston W, Williams IS (1988) Growth of early Archaean crust in the Ancient Gneiss Complex of Swaziland as revealed by single zircon dating. Tectonophysics 161: 271-298

Lagache M (ed) (1988) Thermométrie et barométrie géologiques. Soc Fr Minéral Cristallogr, Paris, 663 pp

Lamb W, Valley JW (1984) Metamorphism of reduced granulites in low-CO_2 vapor-free environment. Nature 312: 56-58

Lamb WM, Valley JM, Brown PE (1987) Post-metamorphic CO_2-rich fluid inclusions in granulites. Contrib Mineral Petrol 96: 485-495

Lancelot JR, Vitrac A, Allègre CJ (1976) Uranium and lead isotopic dating with grain by grain zircon analysis: a study of complex geological history with a single rock. Earth Planet Sci Lett 29: 357-366

Lee HY, Ganguly J (1988) Equilibrium compositions of co-existing garnet and orthopyroxene; experimental determinations in the system FeO-MgO-Al2O3-SiO2, and applications. J Petrol 29: 93-113

Le Breton N, Thompson AB (1988) Fluid absent (dehydration) melting of biotite in metapelites in the early stages of crustal antexis. Contrib Mineral Petrol 99: 226-237

Leake BE (1964) The chemical distinction between ortho- and para-amphibolite. J Petrol 5: 238-254

Lindsey DH (1983) Pyroxene thermometry. Am Mineral 68: 477-493

Lister GS, Snoke AW (1984) S-C mylonites. J Struct Geol 6: 617-638

Lister GS, Williams PF (1983) The partitioning of deformation in flowing rock masses. Tectonophysics 92: 1-33

Maddock RH (1983) Melt origin of fault generated pseudotachylytes as demonstrated by textures. Geology 11: 105-108

Maddock RH (1986) Partial melting of lithic porphyroclasts in fault generated pseudotachylytes. Neues Jb Miner Abh 155: 1-14

Maddock RH, Grocott J, van Nes M (1987) Vesicles, amygdales and similar structures in fault-generated pseudotachylytes. Lithos 20: 419-432

Marjoribank RW, Rutland RWR, Glen RA, Laing WP (1980) The structure and tectonic evolution of the Broken Hill region Australia. Precambrian Res 5: 311-338

Mason R (1981) Petrology of the metamorphic rocks. Allen and Unwin, London, 254 pp

Mattauer M (1986) Intracontinental subduction, crust-mantle décollement and crustal stacking wedge in the Himalayas and other collision belts. In: Coward MP, Ries AC (eds) Collision tectonics. Geol Soc Spec Publ 19: 37-50

McCulloch MT, Wasserburg GJ (1978) Sm-Nd and Rb-Sr chronology of continental crust formation. Science 200: 1003-1011

McCulloch MT, Black LP (1984) Sm-Nd isotopic systematics of Enderby Land granulites and evidence for the redistribution of Sm and Nd during metamorphism. Earth Planet Sci Lett 71: 46-58

Means WD (1981) The concept of steady state foliation. Tectonophysics 78: 179-199

Means WD, Hobbs BE, Lister GS, Williams PF (1980). Vorticity and non-coaxiality in progressive deformation. J Struct Geol 2: 371-378

Mehnert KR (1968) Migmatites and the origin of granitic rocks. Elsevier, Amsterdam, 405 pp

Meissner R (1989) Rupture, creep, lamellae and crocodiles: happenings in the continental crust. Terra Nova 1: 17-28

Meissner R, Strehlau J (1982) Limits of stress in continental crust and their relation to the depth-frequency relation of shallow earthquakes. Tectonics 1: 73-89

Mezger K, Hanson GN, Bohlen SR (1988) U-Pb systematics of garnet. Dating the growth of garnet in the late Archean Pikwitonei granulite domain at Cauchon and Nata-wahunan Lakes. Contrib Mineral Petrol 101: 136-148

Mezger K, Hanson GN, Bohlen SR (1989) High-precision U-Pb ages of metamorphic rutile: application to the cooling history of high-grade terranes. Earth Planet Sci Lett 96: 106-118

Milisenda C, Liew TC, Hofmann AW, Kröner A (1988) Isotopic mapping of age provinces in Precambrian high-grade terrains: Sri Lanka. J Geol 96: 608-615

Molen van der I (1985) Interlayer material transport during layer-normal shortening. Part 1. The model. Tectonophysics 115: 275-295

Moorbath S, Taylor PN (1986) Geochronology and related isotopic geochemistry of high-grade metamorphic rocks from the lower continental crust. In: Dawson JB, Carswell DA, Wedepohl KH (eds) The nature of the lower continental crust. Geol Soc Lond Spec Publ 24: 211-220

Myers JS (1970) Gneiss types and their significance in the repeatedly deformed and metamorphosed Lewisian Complex of Western Harris, Outer Hebrides. Scott J Geol 6: 186-199

Myers JS (1976) Granitoid sheets, thrusting, and Archean crustal thickening in West Greenland. Geology 4: 265-268

144

Myers JS (1978) Formation of banded gneisses by deformation of igneous rocks. Precambrian Res 6: 43-64

Myers JS (1981) The Fiskenaesset anorthosite complex - a stratigraphic key to the tectonic evolution of the West Greenland gneiss complex 3000-2800 my ago. Geol Soc Aust Spec Publ 7: 351-360

Myers JS, Williams IR (1985) Early Precambrian crustal evolution at Mount Narryer Western Australia. Precambrian Res 27: 153-163

Nesbitt HW, Young GM (1984) Early Proterozoic climates and plate motions inferred from major element chemistry of lutites. Nature 299: 715-717

Newton RC, Hasilton HT (1981) Thermodynamics of the garnet-plagioclase-Al_2SiO_5-quartz geobarometer. In: Newton RC, Navrotsky A, Wood BJ (eds) Thermodynamics of minerals and melts. Springer, Berlin Heidelberg New York, pp 131-147

Newton RC (1983) Geobarometry of high-grade metamorphic rocks. Am J Sci 283A: 1-28

Newton RC (1987) Petrologic aspects of Precambrian granulite facies terrains bearing on their origins. In: Kröner A (ed) Proterozoic lithospheric evolution. Am Geophys Union Geodynamics Series 17: 11-26

Newton RC (1988) Nature and origin of fluids in granulite facies metamorphism. In: Ashwal LD (ed) Workshop on the deep continental crust of South India. LPI Technical Report 88-06 Lunar and Planetary Science Institute, Houston, pp 129-131

Newton RC, Smith JV, Windley BF (1980) Carbonic metamorphism granulites and crustal growth. Nature 288: 45-50

Newton RC, Perkins D (1982) Thermodynamic calibration of geobarometers based on the assemblages garnet-plagioclase-orthopyroxene-(clinopyroxene)-quartz. Am Mineral 67: 203-222

Oliver J (1978) Exploration of the continental basement by seismic reflection profiling. Nature 275: 485-488

Park RG, Tarney J (eds) (1987) Evolution of the Lewisian and comparable Precambrian high grade terrains. Geol Soc Lond Spec Publ 27: 221 pp

Parsons I (ed) (1987) Origins of igneous layering. Reidel, Dordrecht. NATO ASI Series C: Mathematical and Physical Sciences, 196 pp

Passchier CW, Simpson C (1986) Porphyroclast systems as kinematic indicators. J Struct Geol 8: 831-843

Passchier CW (1982) Pseudotachylyte and the development of ultramylonite bands in the Saint-Barthélemy Massif French Pyrenees. J Struct Geol 4: 69-79

Passchier CW (1983) Mylonite-dominated footwall geometry in a shear zone, central Pyrenees. Geol Mag·121: 429-436

Passchier CW (1984) Fluid inclusions associated with the generation of pseudotachylyte and ultramylonite in the French Pyrenees. Bull Mineral 107: 307-315

Passchier CW (1986a) Mylonites in the continental crust and their role as seismic reflectors. Geol Mijnbouw 65: 167-176

Passchier CW (1986b) Flow in natural shear zones. Earth Planet Sci Lett 77: 70-80

Passchier CW (1987) Efficient use of the velocity gradients tensor in flow modelling. Tectonophysics 136: 159-163

Passchier CW (1990) A Mohr circle construction to plot the stretch history of material lines. J Struct Geol 12 (in press)

Perchuk LL, Lavrent'eva IV (1983) Experimental investigation of exchange equilibria in the system cordierite-garnet-biotite. Kinematics and equilibrium in mineral reactions. Springer, Berlin Heidelberg New York, pp 199-239

Perkins D, Newton RC (1981) Charnockite geobarometers based on coexisting garnet-pyroxene-plagioclase-quartz. Nature 292: 144-146

Peucat JJ (1986) Behaviour of Rb-Sr whole rock and U-Pb zircon systems during partial melting as shown in migmatitic gneisses from the St. Malo massif NE Brittany, France. J Geol Soc Lond 143: 875-885

Philpotts AR (1964) Origin of pseudotachylytes. Am J Sci 262: 1008-1035

Platt JP, Vissers RLM (1980) Extensional structures in anisotropic rocks. J Struct Geol 2: 397-410

Platt JP (1983) Progressive refolding in ductile shear zones. J Struct Geol 5: 619-622

Poirier JP (1985) Creep of crystals. Cambridge University Press, Cambridge, 260 pp

Ramsay JG (1967) Folding and fracturing of rocks. McGraw-Hill, 568 pp

Ramsay JG (1980) Shearzone geometry: a review J Struct Geol 2: 83-99

Ramsay JG, Graham RH (1970) Strain variation in shear belts. Can J Earth Sci 7: 786-813

Ramsay JG, Huber MI (1983) The techniques of modern structural geology, vol 1: strain analysis. Academic Press, London, 307 pp

Ramsay JG, Huber MI (1987) The techniques of modern structural geology, vol 2: folds and fractures. Academic Press, London, 700 pp

Rozendaal A (1986) The Gamsberg zinc deposit, Namaqualand District. In: Anhaeusser CR, Maske S (eds) Mineral deposits of southern Africa, vol II. Geol Soc S Afr: 1477-1488

Rudnick RL, Taylor SR (1987) The composition and petrogenesis of the lower crust: a xenolith study. J Geophys Res 92: 13981-14005

Rudnik RL, Fievpe˗ ˗ ˗˗˗˗˗ Geochemistry of intermediate- to high-pressure granulites. In: Vielzeuf D, Vidal P (eds) Granulites and crustal differentiation. Reidel, Dordrecht. NATO ASI Series C: Mathematical and Physical Sciences (in press)

Ryan PJ, Lawrence AL, Lipson RD, Moore JM, Paterson A, Stedman DP, van Zyl D (1986) The Aggeneys base metal sulphide deposits Namaqualand District. In: Anhaeusser CR, Maske S (eds) Mineral deposits of southern Africa, vol II. Geol Soc S Afr: 1447-1474

Sawkins FJ (1986) The recognition of palaeorifting in mid- to late-Proterozoic terranes: implications for the exploration geologist. Trans Geol Soc S Afr 89: 223-232

Schärer U, Krogh TE, Gower CF (1986) Age and evolution of the Grenville Province in eastern Labrador from U-Pb systematics in assessory minerals. Contrib Mineral Petrol 94: 438-451

Schenk V (1984) Petrology of felsic granulites metapelites, metabasics, ultramafics, and metacarbonates from southern Calabria (Italy): prograde metamorphism uplift and cooling of a former lower crust. J Petrol 25: 255-298

Scholz CH (1988) The brittle-plastic transition and the depth of seismic faulting. Geol Rundschau 77: 319-328

Schreyer W (1985) Metamorphism of crustal rocks at mantle depth: high-pressure minerals and mineral assemblages in metapelites. Fortschr Mineral 63: 227-261

Schwerdtner WM, Lumbers SB (1980) Major diapiric structures in the Superior and Grenville Provinces of Ontario. Geol Assoc Canada Spec Paper 20: 149-180

Schwerdtner WM (1982) Salt stocks as natural analogues of Archean gneiss diapirs. Geol Rundschau 71: 370-379

Schoneveld C (1977) A study of some typical inclusion patterns in strongly paracrystalline-rotated garnets. Tectonophysics 39: 453-471

Shaw RD, Stewart AJ, Black LP (1984) The Arunta Inlier: a complex ensialic mobile belt in central Australia, parts 1 and 2. Aust J Earth Sci 31: 445-484.

Sibson RH (1974) Frictional constraints on thrust wrench and normal faults. Nature 249: 542-544

Sibson RH (1975) Generation of pseudotachylyte by ancient seismic faulting. Geophys J R Astr Soc 43: 775-794

Sibson RH (1977a) Fault rocks and fault mechanisms. J Geol Soc London 133: 191-213

Sibson RH (1977b) Kinetic shear resistance fluid pressures and radiation efficiency during seismic faulting. Pure Applied Geophysics 115: 387-399

Sibson RH (1980a) Power dissipation and stress levels on faults in the upper crust. J Geophys Res 85: 6239-624

Sibson RH (1980b) Transient discontinuities in ductile shear zones. J Struct Geol 2: 165-171

Sibson RH (1982) Fault zone models heat flow and the depth distribution of earthquakes in the continental crust of the United States. Seismol Soc Am Bull 72: 151-163

Sibson RH (1983) Continental fault structure and the shallow earthquake source. J Geol Soc London 140: 741-769

Sills J D, Tarney J (1984) Petrogenesis and tectonic significance of amphibolites inter-ayered with metasedimentary gneisses in the Ivrea zone, southern Alps, northern Italy. Tectono-physics 107: 187-206

Simpson C, Schmid SM (1983) An evaluation of criteria to determine the sense of movement in sheared rocks. Geol Soc Am Bull 94: 1281-1288

Smithson SB, Johnson RA, Hurich CA, Fountain DM (1986) Crustal reflections and crustal structure. In: Barazangi M, Brown L (eds) Reflection seismology: the continental crust. Am Geophys Union Geodynamics Series 14: 21-32

Spry A (1986) Metamorphic textures. Pergamon Press, Oxford, 352 pp

Strehlau J (1986) A discussion of the depth extend of rupture in large continental earthquakes. Earthquake Source Mechanics 37: 131-146

Suppe J (1985) Principles of structural geology. Prentice-Hall, New Jersey, 537 pp

Talbot CJ (1970) The minimum strain ellipsoid using deformed quartz veins. Tectonophysics 9: 46-76

Tankard A, Jackson MPA, Eriksson KA, Hobday DK, Hunter DR, Minter WE (1982) Crustal evolution of southern Africa. Springer, Berlin Heidelberg New York, 523 pp

Tarney J, Weaver BL (1987) Geochemistry of the Scourian complex: petrogenesis and tectonic models. In: Park RG, Tarney J (eds) Evolution of the Lewisian and comparable Precam-brian high grade terrains. Geol Soc Spec Publ 27: 45-56

Tarney J, Windley, BF (1977) Chemistry, thermal gradients and evolution of the lower continental crust. J Geol Soc London 134: 153-172

Taylor HP Jr (1980) The effects of assimilation of country rocks by magmas on $^{18}O/^{16}O$ and $^{87}Sr/^{86}Sr$ systematics in igneous rocks. Earth Planet Sci Lett 47: 243-254

Taylor SR, McLennan SM (1985) The continental crust: its composition and evolution. Black-well, Oxford, 312 pp

Thiessen RL, Means WD (1980) Classification of fold interference patterns: a re-examination. J Struct Geol 2: 311-316

Thompson AB (1983) Fluid-absent metamorphism. J Geol Soc London 140: 533-547

Thompson AB (1976) Mineral reactions in pelitic rocks. II Calculation of some P-T-X(Fe-Mg) phase relations. Am J Sci 276: 425-454

Thompson PH (1989) Moderate overthickening of thinned sialic crust and the origin of granitic magmatism and regional metamorphism. Geology 17: 520-523

Tobi AC, Touret J (eds) (1985) The deep Proterozoic crust in the North Atlantic provinces. Reidel, Dordrecht. NATO ASI Series C: Mathematical and physical sciences 158: 603 pp

Touret J (1986) Fluid inclusions in rocks from the lower continental crust. In: Dawson JB, Carswell DA, Hall J, Wedepohl KH (eds) The nature of the lower continental crust. Geol Soc London Spec Publ 24: 161-172

Tullis JT, Yund RA (1987) The brittle-ductile transition in feldspathic rocks. EOS 44: 1464

Tullis JT, Snoke AW, Todd VR (1982) Significance of petrogenesis of mylonitic rocks. Geology 10: 227-230

Urai JL, Means WD, Lister GS (1986) Dynamic recrystallization of minerals. In: Mineral and rock deformation: laboratory studies. The Paterson volume, pp 161-199

Valley JW (1988) Granulite melts and fluids in the deep crust In: Ashwal LD (ed) Workshop on the deep continental crust of South India. LPI Technical Report 88-06 Lunar and Planetary Science Institute, Houston, p 187

Vernon RH (1983) Metamorphic processes; reactions and microstructure development. Allen Unwin, London, 247 pp

Vernon RH (1978) Porphyroblast-matrix microstructural relationships in deformed metamorphic rocks. Geol Rundschau 67: 288-305

Vielzeuf D, Holloway JR (1988) Experimental determination of the fluid-absent melting relations on the pelitic system. Consequences for crustal differentiation. Contrib Mineral Petrol 98: 257-276

Vielzeuf D, Boivin P (1984) An algorithm for the construction of petrogenetic grids - application to some equilibria in granulitic paragneisses. Am J Sci 284: 760-791

Watterson J (1968) Homogeneous deformation of the gneisses of Vesterland, south-west Greenland. Meddelelser om Grønland 175: 72 pp; also Grønl Geol Unders Bull 78

Watts MJ , Williams GD (1979) Fault rocks as indicators of progressive shear deformation in the Guingamp region Brittany. J Struct Geol 1: 323-332

Wenk HR (1978) Are pseudotachylytes products of fracture or fusion? Geology 6: 507-511

Werner CD (1987) Saxonian granulites - igneous or lithogenous. A contribution to the geo-chemical diagnosis of the original rocks in high-metamorphic complexes. In: Gerstenberger H (ed) Contributions to the geology of the Saxonian granulite massif (Sächsisches Granulitgebirge) Zfl-Mitteilungen Nr 133: 221-250

Wetherill GW (1956) Discordant uranium-lead ages. Trans Am Geophys Union 37: 320-326

White SH (1979) Large strain deformation: report on a Tectonic Studies Group discussion meeting held at Imperial College, London. J Struct Geol 1: 333-339

White SH, Bretan PG, Rutter EH (1986) Fault-zone reactivation: kinematics and mechanisms. Phil Trans R Soc London A 317: 81-97

White SH, Burrows SE, Carreras J, Shaw ND, Humphreys FJ (1980) On mylonites in ductile shear zones. J Struct Geol 2: 175-187

Whitehouse MJ (1988) Granulite facies Nd-isotopic homogenization in the Lewisian complex of northwest Scotland. Nature 331: 705-707

Williams PF (1977) Foliation: a review and discussion. Tectonophysics 39: 305-328

Williams PF (1983) Large scale transposition by folding in northern Norway. Geol Rundschau 72: 589-604

Windley B (1981) Precambrian rocks in the light of the plate tectonic concept. In: Kröner A (ed) Precambrian plate tectonics. Elsevier Amsterdam: 1-20

Winkler HG (1976) Petrogenesis of metamorphic rocks, 3rd edn.. Springer, Berlin Heidelberg New York, 320 pp

Wood BJ (1974) The solubility of alumina in orthopyroxene co-existing with garnet. Contrib Mineral Petrol 46: 1-15

Yardley BWD (1989) An introduction to metamorphic petrology. Longman, Essex, 248 pp

Zen E-an (1988) Thermal modelling of stepwise anatexis in a thrust-thickened sialic crust. Trans R Soc Edinburgh. Earth Sci 79: 223-235

10 Index